DOWN TO EARTH

Like Atlas, this child aspires to bear the earth on his shoulders ('Faience de Nevers', France, seventeenth century).

DOWN TO EARTH

Foundations Past and Present: The Invisible Art of the Builder

JEAN KERISEL

Emeritus Professor, Ecole Nationale des Ponts et Chaussées, Paris, France

A.A.BALKEMA / ROTTERDAM / BOSTON / 1987

ISBN 90 6191 688 7

Distributed in USA & Canada by: A.A.Balkema Publishers, P.O.Box 230, Accord, MA 02018

Printed in the Netherlands

Table of contents

Introduction

Myths and religions provide us with numerous accounts of belief in a human being who was born out of and will be regenerated by the Earth. His mother, therefore, must be treated with absolute respect. This in particular lay behind the teachings of the Indian prophet Smohalla: 'It is a sin to wound or to cut into, to tear or scratch our common mother... You ask me to plow the ground! Shall I take a knife and plunge it into my mother's bosom? Then when I die she will not take me to her bosom to rest. You ask me to dig for stone! Shall I dig under her skin for her bones? Then when I die I cannot enter her body to be born again'.

Nevertheless, at the time when man first took to a sedentary existence, the reindeer hunters hollowed the soil to form a trough within the circle of stones placed to secure the walls of their tents and, in their desire for snugness and warmth, left the first mark on 'Mother Earth'. From then on, rooting things in the earth was to become one of the adventures in the early life of man.

This makes it all the more interesting to ask ourselves how, over the ages and in spite of the myths which maintained their hold here and there, our earth has accepted, preserved or rejected her burden of civilizations and how the visions of generations of builders have taken form upon her body.

It was on the soft clay of tablets that man wrote down his first thoughts and the depth of his imprint depended only on him; but the earth herself generates forces of gravity which leave gashes of their own. The marks left by generations of human beings and the imprint of civilizations have come to depend only on the ambitions and the skill of builders. But whereas the forces of gravity retain their essential mystery, the efforts to neutralize them or profit from their effects is a striking pointer to the art of the builder.

It is an art of building that is hidden, an invisible architecture concealed under or behind the visible exterior it makes possible. It is even, as we shall see, the very key to the construction of religious buildings: a vault, as Leonardo da Vinci remarked, has two weak points which both work for its collapse but which can be transformed into a strength; we shall show that it is the invisible architecture under the ordinary or flying buttress which gives strength to the vault and enables it to reach towards the heavens.

Historians and writers search for the roots of ethnic groups, families and individuals; but it is also interesting to search for those of major human constructions and to admire the sensitivity of certain builders in their relationship with the earth..

These ancient remains are often enigmatic, as the philosopher Alain notes: 'A man who lies down in the grass writes his form in it, as a dog or a hare does, and since man thinks and tosses around in his thoughts, I may say that man writes his thoughts in his bed of grass. In fact, it is not easy to read this writing; that is why all of the plastic arts are enigmatic. Man himself is an enigma in motion; his questions never stay asked; whereas the mould, the footprint, and by natural extension, the statue itself, like the vaults, the arches, the temples with which man records his own passing, remain immobile and fix a moment of man's life'.

An enigma for two reasons: for one thing, ancient civilizations made use of sun-baked bricks, with the result that the glorious remains of their monuments are intermingled with their ruins; it required all the ingenuity and patience of Koldewey to brave this confusion and unravel the layout of Babylon.

Secondly, the traces left behind often escape our notice because the civilizations of the past made but a shallow mark upon the soil: it took Schliemann and the archaeologists who followed him several decades to uncover at Hissarlik the seven superposed towns of Troy, which condensed barely two thousand years of human history into a layer no more than twenty metres thick, whereas the powerful erosion of the Colorado River biting into the Grand Canyon offers the spectacle, as we contemplate its 1600 metre high cliffs, of over two thousand million years of the earth's history.

The archaeologist is interested in the superposition of strata, which give him a key to the succession of civilizations; our investigation, on the other hand, is concerned with the general form of the non-visible architecture. For example, the order of the Aztec dynasties is of much less interest to us than the deeply buried foundations of the Templo Mayor, which the Aztecs built on the Lake Texcoco, where Mexico City now stands. These foundations, planted deep within the sediments of the lake discovered under the historic quarter of Mexico City (Matos Moctezuma 1982) escaped the destructive rage of Cortes and his determination to level everything to the ground; they reveal that the Aztecs, in attempting the impossible – to build a monumental temple on a lake – recognized the need for gentleness when building on a soft soil and knew that they had to let that soil rest from time to time before proceeding further.

Man has left countless traces of his energy. We shall choose among those which suggest comparisons, mark contrasts and emphasize the continuity between past and present.

These traces we shall seek mainly in the Mediterranean basin, which was at least a thousand years ahead of the other civilizations and stands as a witness to the achievements of the Judeo-Christian civilization. Some exceptional documentation offered by Professor Lu Zhao Jun will also reveal that the Chinese of the first millennium were also masters in the art of foundations.

The Chaldean ziggurats, the pyramids of Egypt and the Americas, the stoas of Greece, the Roman roads, the walls of Attalus, the Chinese monuments of the first millennium, the Italian towers built at the beginning of the second millennium and the great cathedrals are just some of the outstanding moments in the adventure of mankind through which we shall try to grasp the intelligence of man in perfecting the structures which he anchored in the ground.

Though mistakes were made and many ancient skills forgotten, we have much to learn from the experience of the past, which inspired our modern architecture in the pre-industrial and industrial periods.

When a major discovery is made, as F. Russo has remarked, we tend to give all the credit to the person who apparently played the leading role and to date the advance at a certain point in time. In many cases, however, this man has simply made use of existing knowledge connected with the discovery and it would be unwise for us to disregard the legacy of previous centuries.

The study of foundations now forms part of the discipline known as soil mechanics. This science deals with the laws governing the equilibrium, deformation and rupture of the earth's crust, on land or under water, as a result of the variety of loads placed upon it by man or of the unloading occasioned by his cutting into slopes; it is therefore not only the science of foundations on the earth's crust but also that of the inclusions needed to strengthen, buttress and retain this soil, of deep incisions and of subterranean penetration.

Initially, at the beginning of industrial civilization, soil mechanics formed part of the general art of building. It gradually evolved into a separate discipline and later extended its scope to become the science which studies certain forms of construction (tombs, tumuli, ziggurats, mastabas, dolmens, pyramids) and the building of dykes, dams and in fact any work made of earth (soil, silts, stones and rocks), so much so that we shall often have to deal with architecture in general, that is, with the art of building these structures. More recently, it has further broadened its field to play a part in the protection of the environment by studying the rational exploitation of natural resources, such as quarries and mineral deposits, and the return of waste matter to the earth.

In the last part of the book we shall place ourselves in the situation of an observer seeking the traces left by our present civilization after a lapse of time equal to that from which we contemplate the traces left by the first sedentary civilizations. For this archaeologist of the year 14000, separated from us by a mere instant of geological time, we shall examine the remains of a past that is our present. This attempt at futurology draws its inspiration from the continuity of a past in the course of which many imprints have slowly and inexorably disappeared at a pace imperceptible to man while others have vanished suddenly, erased by earthquakes or some other disaster. And in this final section, which could have been entitled 'From the visible to the invisible', we shall ask ourselves what, for our observer of the year 14000, will remain of the various great cities threatened with slow or sudden extinction – Venice, Bangkok, London, Hamburg, San Francisco, Tokyo – and consider a few of the problems which have already aroused public concern.

Part One
The past

* 1 *

The earth, animal life and mankind

The building instinct of the first sedentary human beings developed from a certain intuition acquired by the hominids during their first attempts to walk upright and, even further back in time, from the experience of anthropoid apes and mammals of the tertiary as they advanced with difficulty on spongy soils, like those of the big deltas of Mesopotamia, the Nile and the Indus, where the first civilizations developed. For the very early period, the illustrations and text of the book 'Sulle orme dei Dinausori' (On the tracks of dinosaurs – Bonaparte et al., 1984) are fascinating: the foremost palaeontologists of the present day show how the dinosaurs left their mark in every continent (Photo 1). While man has lorded the earth for no more than a few millennia, the reign of the dinosaurs lasted some 150 million years, during which the heaviest animals ever known left their deep imprints in the soil. Thrilling too are the miraculously preserved footprints discovered by Dr. Mary Leakey (1979) (Photo 2), which represent two hominids walking side by side about three and a half million years ago and leaving the trace of their presence in the film of ash spewed from a volcano in the Laetoli plain of Tanzania. How tenderly we watch the first faltering steps of a child. And how could we fail to be moved by this glimpse of the creatures of Laetoli, who were perhaps the first ones to walk upright. The fine layer of ash that covered the clay moulding their footsteps has bequeathed to us a miraculous message, delivered intact after several million years. These footprints show that the human foot had completed its evolutionary process from that of a monkey (Figure 1), making upright motion possible. In hunting for prey or escaping from a predator, man (Figure 2) abandoned the taligrade or plantigrade postures to become digitigrade: the metatarsal bar BC and the hallux c form in the soil what amounts to a spring-board T. The visible element is the springing of the foot; what is invisible is the support provided by the soil which can give way (along the dotted line L) if necessary. The poet Valéry said: 'Only the body knows and only the body which responds without thought to obstacles possesses the highest knowledge'. Here the body has already found the principle of a mechanical coupling and the idea of the progressive rupture of a support.

Photo 1. Toro Toro (Bolivia). The gigantic footprints of a dinosaur about 190 millions years old. The man in the background gives an idea of their size. By permission Centro Studi Richerche Ligabue, Venice.

Photo 2. Footprints of hominids in the Laetoli plain in Tanzania, discovered by Dr. Mary Leakey (about 3.6 million years old).

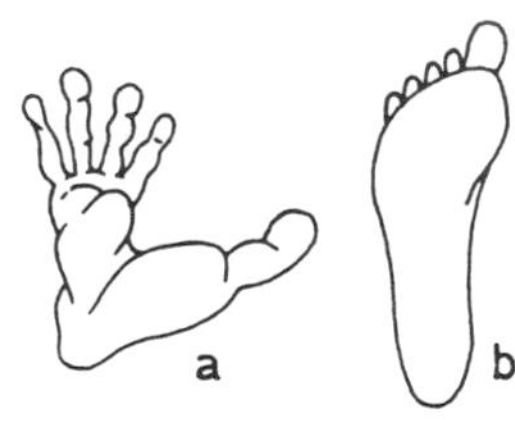

Figure 1. Foot a) of the monkey b) of the human being

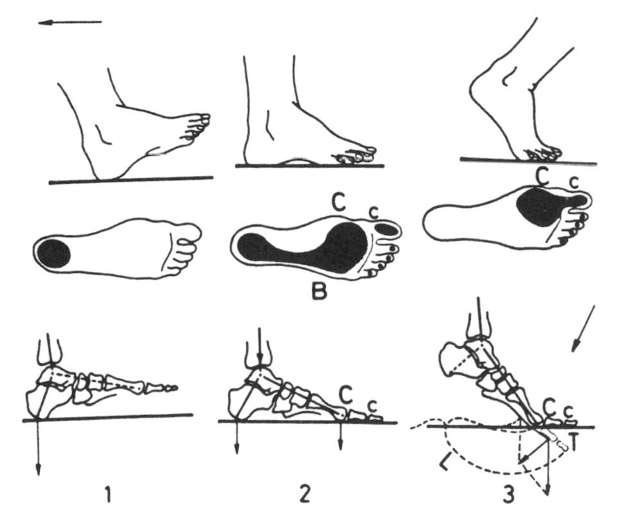

Figure 2. The three upright postures of man.
1: Taligrade 2: Plantigrade 3: Digitigrade

Figure 3. The stone tool, precursor of the rock bit used for oil drilling, wielded by prehistoric man. Drawing by P. Vaughan published in 'La Recherche' (1983).

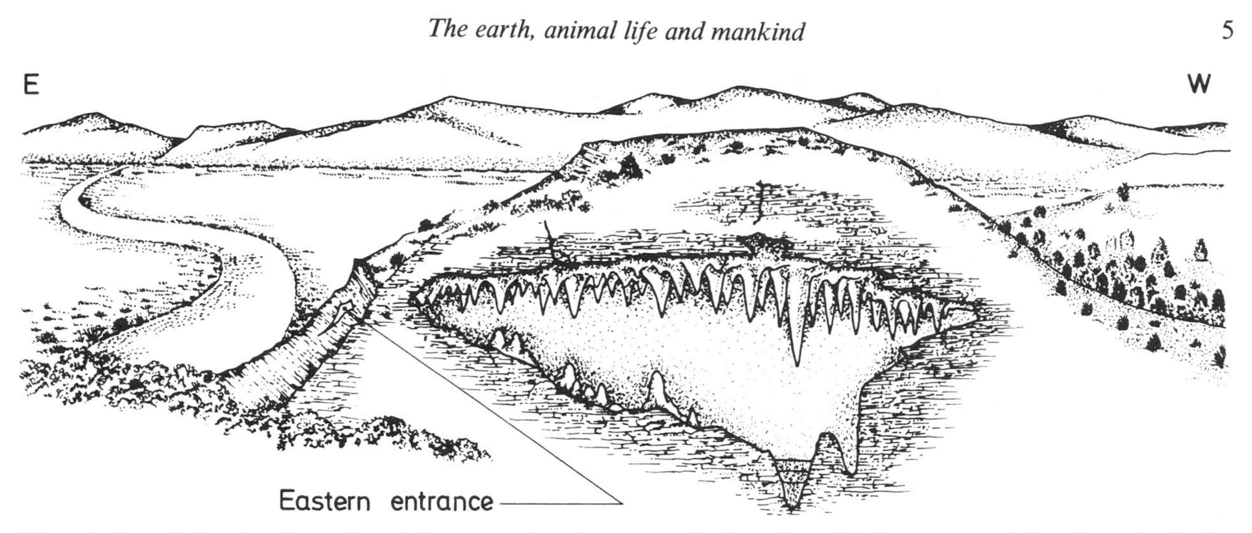

Figure 4. Cave of Choukoutien at time of first occupation. An entrance has been formed at the eastern end through erosion by the Choukou river. The cave was inhabited by Homo Pekinensis for 230000 years (from Wu Rukan and Ling Shenglong 1982).

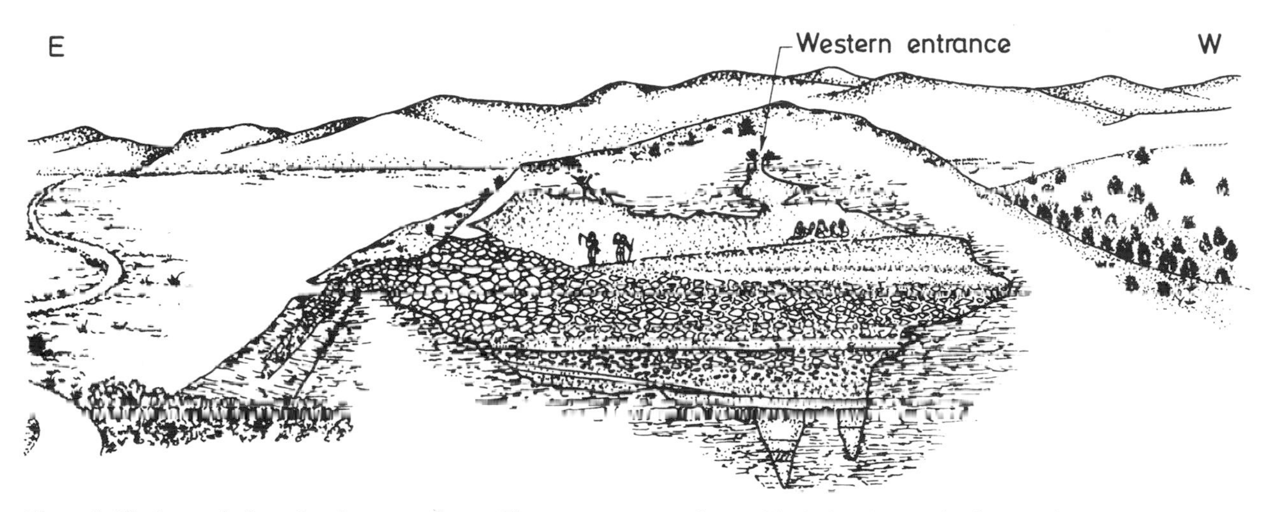

Figure 5. Final stage before abandonment of cave. The eastern entrance is now blocked and access is via a crevice to the west. The cave has been filled by the detritus of generations of Peking Man (from Wu Rukan and Ling Shenglong 1982).

Long before man, the elephants used to test this progressive rupture before crossing doubtful fords or muddy swamps by poking their trunks deep into the soil or by kneeling.

The hominid therefore bore within himself the harvest of a long evolution which, much later, would teach prehistoric man to fashion a variety of stone tools with which he scratched the ground with the first forms necessary to his existence. The gestures of twisting and striking with his tool (Figure 3) foreshadow the movements of the present-day rock bits and rockdrills used in boring for oil. These implements enabled him to dig ditches to keep his shelter dry and to arrange the caves in which he lived during the Palaeolothic period, which accounts for 98% of human life on earth. In China, the cave of Choukoutien has yielded indisputable evidence of man's growing knowledge of rocks. This cave was continuously inhabited by Peking Man for 230000 years, from 560000 to 230000 BP, that is, for twenty times longer than the period stretching from the end of the Palaeolithic (10000 BP) up to the present day.

The cave is about forty kilometres southwest of Peking. Excavation started in 1920 and has only recently been completed. The Chinese geologists Wu Rukan and Ling Shenglong (1982) have succeeded in retracing the evolution of this gigantic karstic

Photo 3. Cave at Font de Gaume. Shape of a barrel vault symbolizing the effect of arching (reproduction authorized by the Conservation des Monuments Historiques pour l'Aquitaine).

cavity, the formation of which began some five million years ago. It developed gradually within a hillside but had no outlet. Later (Figure 4), as a result of erosion of the eastern flank by a river flowing alongside the hill, an opening was formed about 460 000 years ago and Homo Erectus Pekinensis settled in that spacious shelter, at one time over 140 m long. Finally, around 300000 years ago, a large rockfall in the eastern part forced the cave-dwellers to take refuge in the western part, which had by then become accessible (Figure 5) through a crevice in the limestone. By the time it was abandoned, around 230000 BP, the cave had become filled with debris from minor rockfalls and the detritus of human occupation. It is possible to distinguish ten separate layers in this debris (Teilhard de Chardin helped with its classification in 1929) which includes the remains of forty individuals, less interesting for our purposes however than the tools they used. The tools of the first period (460000 to 420000 BP) are relatively large and shaped from comparatively weak sandstone. In contrast, those from the final period (360000 to 230000 BP) are smaller and made of quartz or occasionally of flint. It is now possible for us to imagine how, with these hard pointed stones, prehistoric man managed to dig, arrange his cave, widen the narrow western entry and so forth.

In brief, Peking Man, in seeking rocks, testing their hardness and examining how they split was, by necessity, a student of geology and soil mechanics.

Still more recently, prehistoric man, now beginning to express his thoughts in rock engravings, appears to have sensed by intuition the vault-like manner in which the stresses worked upon the rock mass above his head. How else can one explain the series of parallel lines in the form of an arch (Photo 3) chiselled by a home sapiens some twenty to thirty thousand years ago in the natural roof of the cave at Font de Gaume, in the Dordogne region of France? They seem to prefigure the steel linings now used for tunnels cut through loose ground.

But this was simply a question of instinct. Both animals and humans (before they became sedentary) appear to have given thought to improving the ground on which they dwelt. According to certain specialists in prehistory, the monkeys, like beavers, knew how to force piles into the soil in order to construct resting platforms. Isernia Man himself (around 700000 BP), whose skeleton was recently discovered in southern Italy, used to spend some of his time in marshy areas and was wont to improve certain crossing places by inserting the long bones of bison, rhinoceros and elephants (Salvatori 1984).

* 2 *

The ziggurat: Invention in ancient Mesopotamia

When man gave up the nomadic life, he often made his home in the deltas of great rivers. Though this had its advantages for trading, the weak bearing capacity of the soil caused him all sorts of difficulties for his settlements. The habitations were very light, made of reeds and branches – an aquatic architecture that was, so to speak, the harvest of the soil of the delta which, in Mesopotamia, was soft alluvium borne down towards the Persian Gulf by the two great rivers, the Tigris and the Euphrates (Figure 6). These rivers carried so much silt to the almost level plain that their beds rose with the result that, at flood time, they overflowed their banks and spread the alluvium over the surrounding land.

The Greek geographer Strabo (c. 60 B.C.–c. 15 A.D.) describes the plain as follows: 'The soil is deep and soft; it yields so easily that the trenches and canals are choked or silted up and the plains near the coast form lakes and marshes filled with reeds'.

For a long time, it was thought that this alluvium was helping to push the coastline southwards, filling in that part of the Persian Gulf. In 1952, however, two geologists, Lees and Falcon, were able to show that the position of the coast had hardly changed in historical times: the amount of alluvium carried by the two powerful rivers (about 60 cm per 100 years) was almost entirely counterbalanced by the settlement of the overloaded soil.

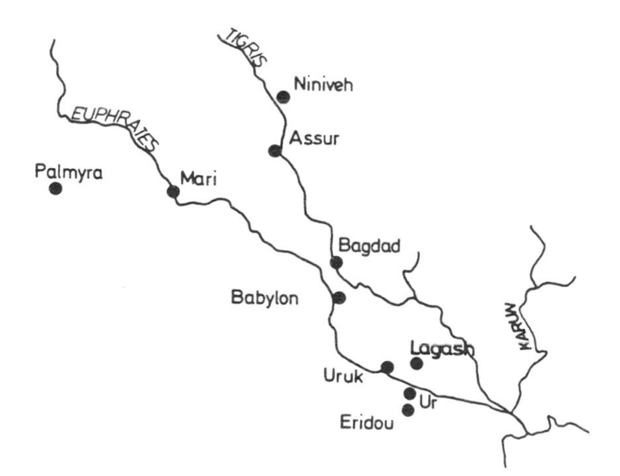

Figure 6. Map of Ancient Mesopotamia.

Nevertheless, it was on this discouraging terrain that the Sumerians wanted to express their religious faith by building ziggurats, high massive terraces on top of which they erected a temple.

In 1948, the excavations by Safar and Lloyd at Eridu revealed that, in this city considered by the Sumerians as the oldest in the world, the ziggurat of temple I (Photo 4) covered no less than eighteen successive levels of occupation, the first of which, at a depth of about 12 metres, dates back almost to the beginning of the Ubaïd period (4000 B.C.) while the uppermost ones relate to the end of the Uruk-Ubaïd period (up to 3000 B.C.).

The last ziggurat, the ziggurat of Etemenanki in Babylon, was built by Nebuchadnezzar II around 600 B.C. Thus, for more than three millennia, ziggurats were built in Mesopotamia, in most cases on a hostile soil.

With little stone available, their builders made use of sun-baked bricks laid out in successive courses. They tried to produce sharp almost vertical edges that would stand out against the flat expanse of land and water. But the architects were rapidly faced with a problem similar to what modern engineers call 'quick sands': the alluvium soon yielded under the weight while, at the same time, the base of the ziggurat spread outwards. This explains the large number of successive levels of occupation, the earliest ones being capped only with simple chapels and subsequently used as filling.

Little by little, the settlement and spreading (Figure 7) diminished until it eventually became possible to build the temenos, the sacred enclosure sur-

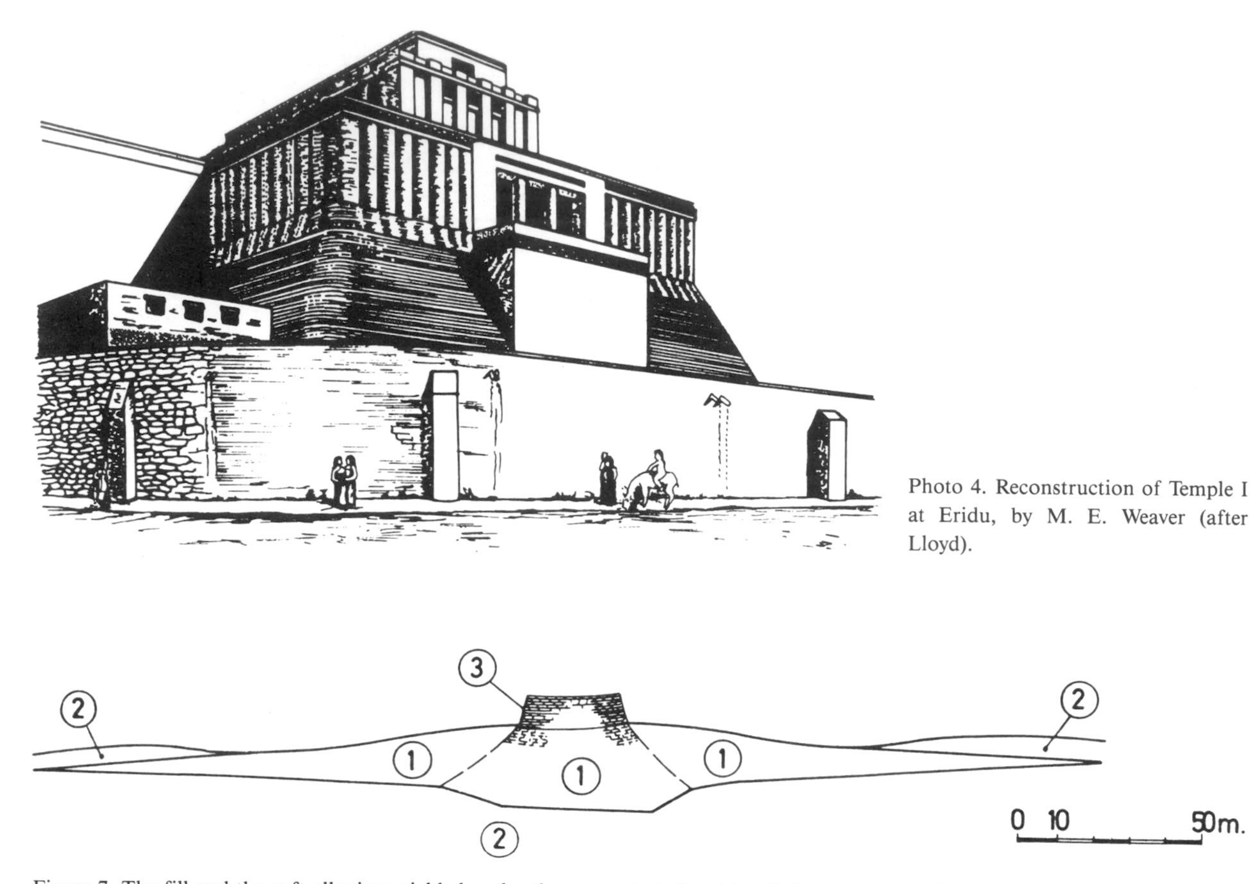

Photo 4. Reconstruction of Temple I at Eridu, by M. E. Weaver (after Lloyd).

Figure 7. The fill and the soft alluvium yielded under the concentrated weight of the temenos: author's interpretation, based on his experience with the Arzal dam (Kerisel 1970).
(1) Fill (2) soft soil (3) Temenos
Probable slip line ______ ______ ______.

rounded by retaining walls, on which stands the platform bearing the temple with its portico. Photo 4 represents an attempted reconstruction of the layout imagined after studying the remains of other ziggurats and temples.

Constructions on this scale made it necessary to halt, wait and then resume the building process many times. Around 2100 B.C., however, the Sumerians, now free from the Akkadian yoke, made astonishing progress. Wooley (1939) has shown that, in building the ziggurat of Ur-Namu (61×46 m at the base), they placed thick layers of woven reeds every six to eight courses of sun-dried brickwork. Later on, under the Kassite dynasty (15th to 13th century B.C.), the ziggurat at Aqar Quf, seat of the Kassite capital to the west of Bagdad, was built of sun-dried bricks with, every eight or nine courses, an 8 cm layer of sand containing mats of woven reeds and powerful cables of plant tissue which together absorbed the horizontal thrust that tended to split the mass of the ziggurat. Evenly spaced from bottom to top these reinforcements permitted a graceful form with nearly sheer edges since the layer of sand acted as a drain, sucking water from the bricks and thus increasing their density so that further courses could be added without crushing of the bricks.

A remarkable invention: indeed, no other period, with the exception of the nineteenth century, engendered as many technical innovations as the civilization of Mesopotamia. Photos 5 and 6 show the remains of this ziggurat and a reed mat. The diagrams in Figure 8 represent a comparison between the present and primitive states of the ziggurat, according to the Turin archaeologists: one can see how greatly time has eroded this monument which nevertheless remains one of the finest examples of this type of Mesopotamian construction.

Photo 5. The ziggurat at Aqar Quf.

Photo 6. A reed mat. Photo D. Parry.

After the period of Assyrian domination and during the last Mesopotamian monarchy (6th century B.C.), Nebuchadrezzar II gave new life to the Babylon of Hammurabi's time in the 19th century B.C.

Thanks to the patient and skillful excavations of the German archaeologist Koldewey at the beginning of this century, we can now trace with confidence the plan of the great city (Figure 9) before it was besieged by Cyrus in 562 B.C. The city was built on both sides of the Euphrates alongside quays connected by one of the largest bridges in the memory of man. The Procession Way started from the Ishtar Gate and ran along the east side of Nebuchadrezzar's Palace, the ziggurat of Etemenanki and the temple of Marduk. Figure 10 (Koldewey) represents a vertical section of the Procession Way and Ishtar Gate, whose front view and decorated foundations are illustrated: in all probability, these decorations gradually sank to a depth of 15 m during the construction of the famous gate, a model of which is in the Staatliche Museen of East Berlin.

The Procession Way was paved with limestone slabs roughly one metre square. One of these slabs has the following cuneiform inscription: 'I am Nebuchadrezzar, King of Babylon, son of Nabopolassar, King of Babylon. I have paved the Babel Street with slabs from Shadu for the Procession of the great God Marduk'.

Very little remains of the temple of Marduk and the ziggurat. We can discern remnants of the outer wall of the temenos, which surrounded an area of 500 m^2. The reconstruction of that 75 m high ziggurat with its six terraces has been imagined by Koldewey (Photo 7). It was built of sun-baked bricks, like the huge fortifications of the city with their double enclosure walls, the inner one of which was no less than 6.5 m thick. In erecting the massive ziggurat the Sumerians probably employed the strenghthening techniques we have just described, which turned out to be all the more effective in that the greater part of the ziggurat was founded upon the same site as the one constructed by Hammurabi and completely destroyed together with the city by the bloodthirsty Sennacherib, king

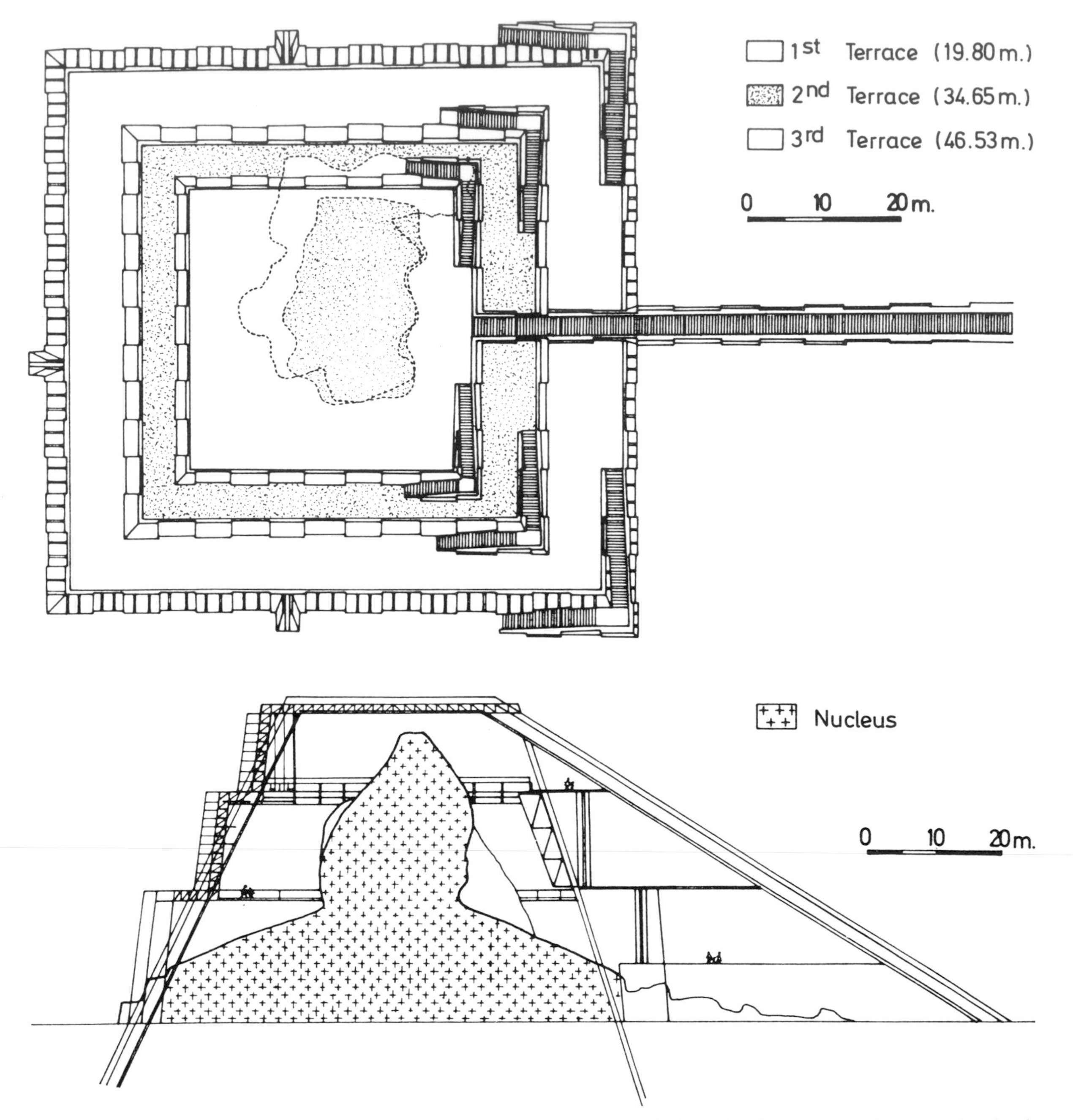

Figure 8. Original form of the ziggurat of Aqar Quf as deduced by Turin archaeologists. View from above and cross-section showing the great ramp and proposals for the reconstruction of the terrace using a substructure of ironwork. From Gullini 1985.

of Nineveh. A cuneiform inscription bears witness to his fury: 'I have destroyed the city right down to its foundations...I have followed this up by a flood...I have ordered the materials extracted from the lower foundations to be thrown into the Euphrates so that they may be carried to the sea'.

Everything in Babylon recalls the grandeur of the ideas with which Genesis credited the Sumerians: 'Let us build us a city and a tower whose top may reach unto heaven and...make us a name'. Since the ziggurat was located on Babel Street it is by no means impossible that it was in fact the Tower of Babel, which could have been built during the relatively short reigns of Nebuchadrezzar and his fa-

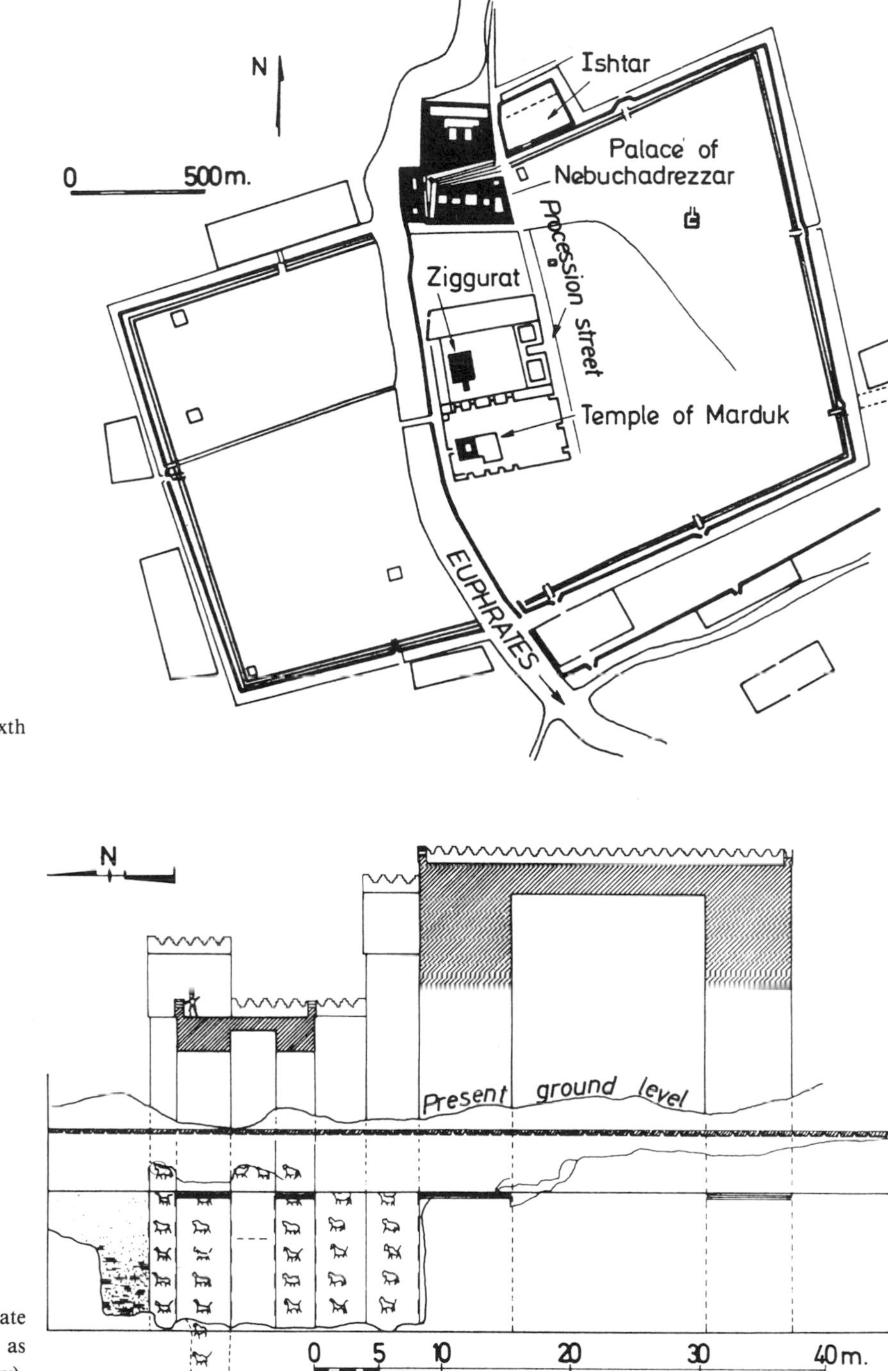

Figure 9. Plan of Babylon in the sixth century B.C. (from Hawkes 1974)

Figure 10. Drawing of the Ishtar Gate and its decorated foundations, as seen from the west (from Koldewey).

ther using Sumerian techniques on a soil that had been preloaded during the reign of Hammurabi. Here there is nothing like the eighteen levels of occupation found at Eridu.

The ziggurat and the city fascinated both Herodotus and its conqueror Cyrus. The latter did not harm the ziggurat but it was partly destroyed by another Persian King and left in ruins until restored for a while by Alexander the Great.

At a time when one of the acknowledged objectives of industrial research in the technologically advanced countries is to develop composite materials, it is interesting to note how this technique is anticipated by the Sumerian invention of reinforced

Photo 7. The model of sixth century B.C. Babylon in the Staatliche Museen of East Berlin. The ziggurat of Etemenanki is on the left and the bridge over the Euphrates in the foreground.

courses of sun-baked bricks. This original idea of combining complementary materials was soon lost: for centuries no more heavy buildings would be constructed on soft soils except by mistake and it was not until near the end of the Aztec civilization that something similar was attempted once again. In 1329, the Aztecs decided to build their capital Tenochtitlan on a shallow lake with a dangerous subsoil of aeolian clay containing ten times more water than solid matter. It was on an artificial island, known as Dog's Island, that they undertook to erect the Huey Teocalli or Templo Mayor, which was 36 metres high when Hernan Cortes arrived at the beginning of the sixteenth century. A worthy disciple of Sennacherib, he razed the entire pyramid to the ground. Archaeologists (Matos Moctezuma 1982) have recently discovered its foundations beneath Mexico City, not far from the Zocalo (Figure 11,) near where the Cathedral and National Palace now stand. The map shows that these two monuments, like the Huey Teocalli, were built on the lacustrian filling of Dog's Island, and this explains why they are not stable even today. The latest research concerning the base of the Templo Mayor (Mazari 1985) reveals that, along with the island on which it stood, it had sunk about 5.6 m by the beginning of this century (Figure 13), though most of this settlement had already occurred by the time of the Temple's demolition by Cortes. The Aztecs were unaware of the Sumerian techniques of the

Figure 11. Tenochtitlan and Mexico City. The broken line shows the probable limits of the lacustrian filling (Dog's Island) on which the Templo Mayor (TM in the diagram) and later the Cathedral and the National Palace were constructed (from Mazari et al.).

second millennium B.C. yet, as at Eridu, they made lengthy pauses during the period of construction, a fact which helps to explain why the pyramid was built in six stages (Figure 12).

Since the beginning of this century, the base of the Templo Mayor has settled, along with the central part of Mexico City, a further 5 to 6 metres owing to imprudent pumping from the water table (Carillo 1948). All in all, the Templo Mayor, like the Temple I erected at Eridu nearly four thousand years ago, has sunk about 12 metres.

These foundations have had a curious destiny. They existed during the heroic age of ancient Tenochtitlan and supported one of the masterpieces of Aztec art. Floating on loose soil, not far from what is now the square of 'the Three Cultures' which commemorates the mixing of the races that has produced present-day Mexico, they are now supporting parts of the modern capital.

After the Aztecs, people stopped building religious structures on such hostile soil. It was not until the nineteenth century that a few more attempts were made, such as the construction in 1820, at the mouth of the Adour in France, of military buildings on sand piles inserted into the soft soil (Kerisel 1956). In the present century the Sumerian ideas have been revived on a vast scale with the technique of inserting steel wires and, nowadays, mats or grids made of polymer fibres (Figure 14) into the filling.

In many cases, modern techniques do not make a clean break with the past; inventions are the intelligent adaptation of an artisanal process which gives substance to an idea. All too often man is under the impression that he is creating ex nihilo: were the Sumerians themselves really inventors? Their ancestors had long mixed straw with silt to strengthen their sun-baked bricks, and perhaps these ancestors themselves, in creating such composite materials, had simply observed the life of plants and noticed how trees strengthen the soil on which they stand with their roots.

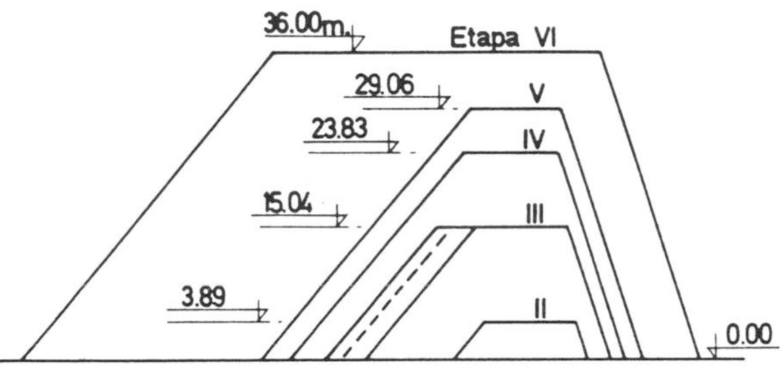

Figure 12. The six stages in the construction of the Templo Mayor (from Mazari et al.).

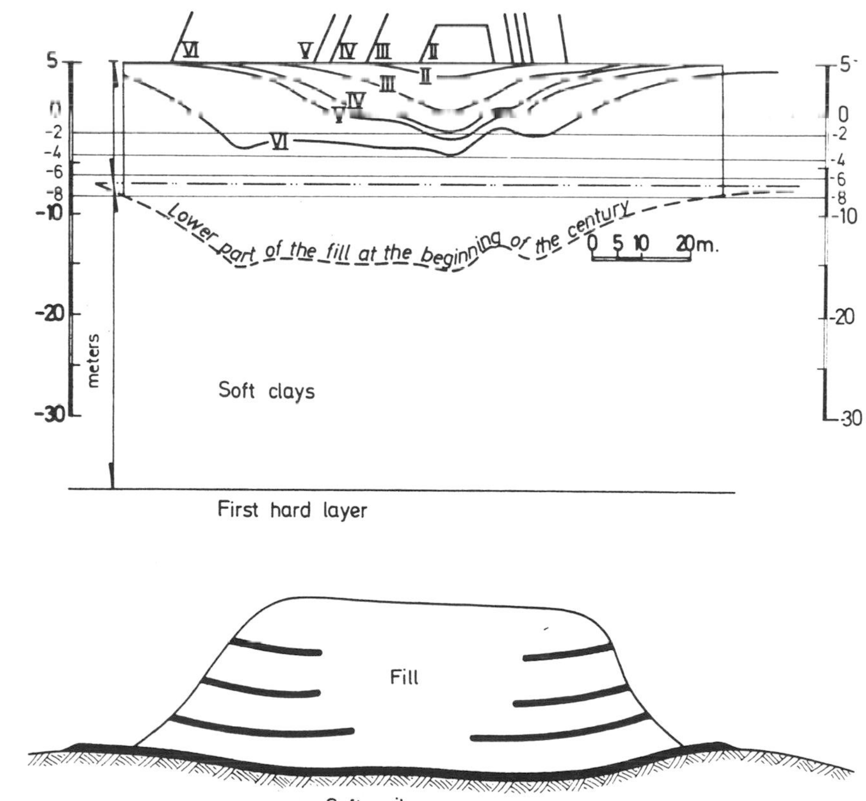

Figure 13. Settlement of the Dog's Island fill and of the base of the Templo Mayor at the beginning of the century (from Mazari et al.)

Figure 14. Adaptation of Sumerian techniques in the present century. Artificial fibre mats (or steel rods) to control settlement and outward thrust and to strengthen the slopes.

* 3 *

The invisible and the visible in the Egyptian pyramids

Several centuries separate the very first ziggurats in Lower Mesopotamia (Eridu) from the pyramids of Egypt. Around 2750 B.C., it became possible to construct these stone monuments whose present state, after five thousand years in the desert, still amazes us. The conditions were ripe:

– the powerful River Nile linked the East Bank quarries of Upper Egypt (Aswan granite) and Lower Egypt (Tura limestone) to the West Bank of Lower Egypt where, at the edge of the Libyan Desert, on a site with a good bearing capacity, the Pharaohs built the pyramids (Figure 15),

– the discovery, in the third millennium B.C., of how to smelt copper made it possible to mould the copper saws used for cutting the stone,

– an autocratic regime in the service of a religion enabled the first Pharaohs to impose the extremely hard work involved on their people,

– lastly the 3rd dynasty saw the birth of an architect of genius, Imhotep, who found effective solutions to the construction problems of the first great pyramid, that of Saqqarah.

In exploring our theme, from the invisible to the visible in architecture, we shall concentrate on two aspects of the design of the great pyramids – their internal structure and the burial chambers they housed – and on the seven major pyramids of Egypt, (Figure 24) which were all erected within a brief period of time (2750-2500 B.C.) in the following order:

– Saqqarah: Stepped Pyramid; beginning of 3rd dynasty (Zoser)

– Meidum: Stepped Pyramid; end of 3rd dynasty (transformed into a true pyramid by Snefrou at the beginning of the 4th dynasty) (Huni)

– Dahshur South: the 'Bent Pyramid'; beginning of 4th dynasty (Snefrou)

– Dahshur North: the first true pyramid; beginning of 4th dynasty (Snefrou)

– Cheops: True pyramid; 4th dynasty (Cheops)

– Chefren: True pyramid; 4th dynasty (Chefren)

– Mycerinus: True pyramid; 4th dynasty (Mycerinus)

The last three form the famous triad of Giza.

BEFORE THE PYRAMIDS

Not much is known about the royal tombs of the 1st and 2nd dynasties. A few of them to the North of Saqqarah were excavated by Emery (1949) between 1935 and 1956. The most ancient one took the form of a shallow trench (19.10 m long, 2.90 m wide and 1.35 m deep) which was divided into five compartments with a ceiling of wood. Over this was placed a much larger structure built of sun-dried bricks and divided into 27 chambers (Figure 16), which were no doubt meant for the jars of food and wine needed by the deceased after his earthly death. The exact height of the walls and form of roof are a matter of conjecture. This is also true for Abydos in Upper Egypt where, probably owing to the unification of Upper and Lower Egypt, similar tombs are to be found. One plausible theory is that the 27 upper chambers were filled with sand so as to overflow the top of the walls, giving the monument a convex roof stopping at the edge of the outside walls which, for greater stability, were provided with thick external buttresses. In short, it was like a low ziggurat (probably 2.50 – 3.00 m high) but with a structure resembling a set of sand-filled silos.

The pilgrims of the Middle Ages who plucked up the courage to visit the monuments of Egypt gave

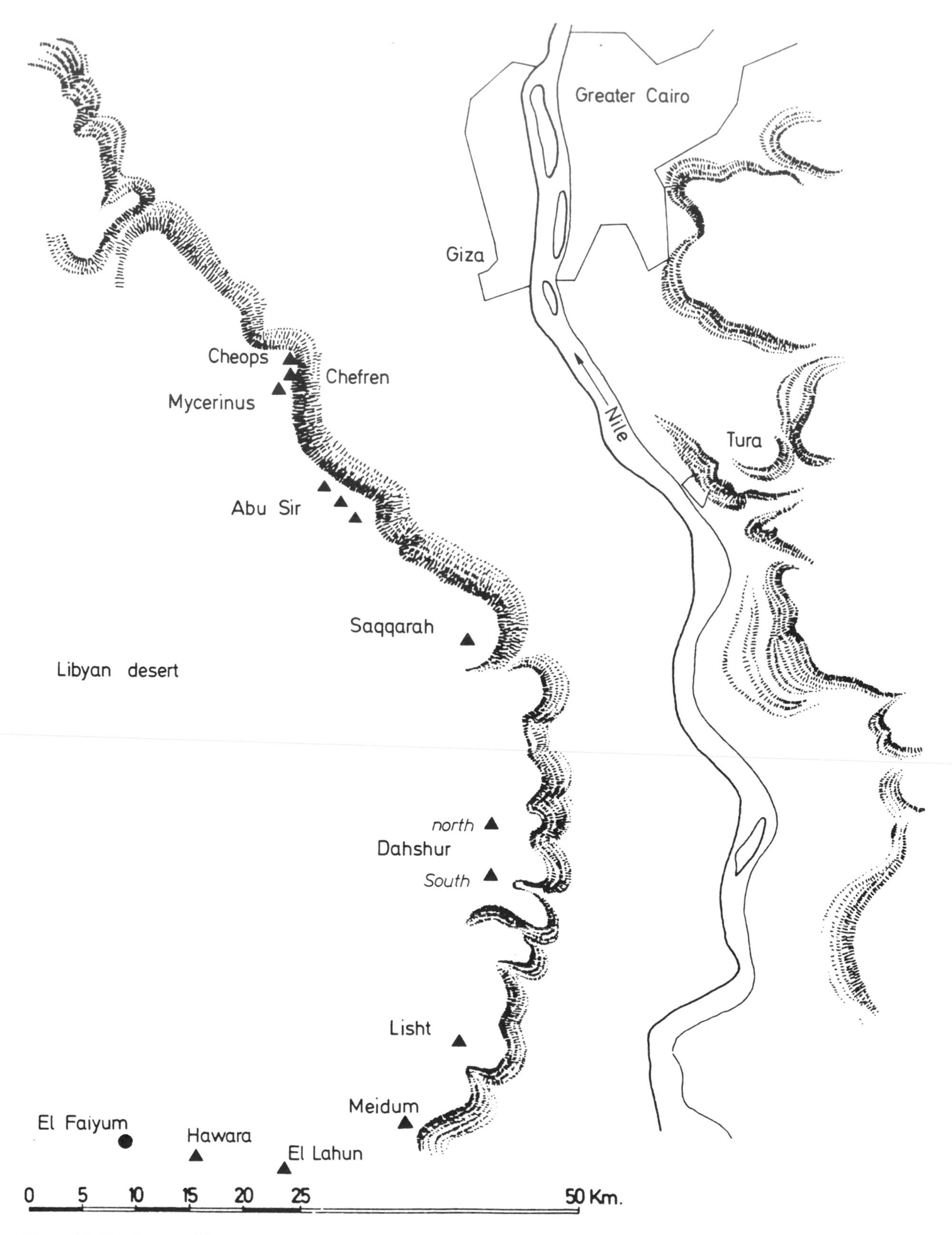

Figure 15. Sketch map of Lower Egypt.

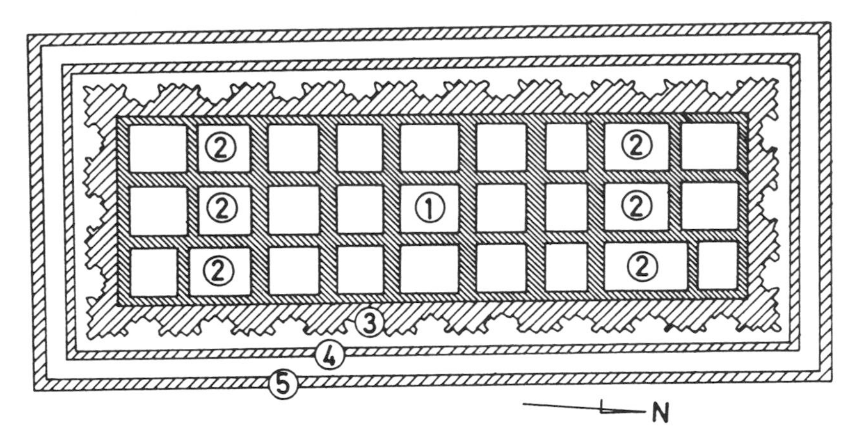

Figure 16. Mastaba from the reign of Aha, second Pharaoh of Upper and Lower Egypt (1st dynasty). (after Emery)
1. Burial chamber 2. Sand 3. External brickwork of the superstructure 4 and 5. Two walls separated by paving stones set in a clay mortar.

credence to a tradition relating that the pyramids had been built by Joseph, son of Jacob, to store the wheat harvested in good years, and called them the 'granaries of Joseph'. The very first pharaonic tombs were indeed in the form of silos... but filled with sand and overlying a tomb.

During this transitional period, we are still far from the pyramidal form. In Egyptian hieroglyphs a true pyramid was represented by a sign like the Greek capital letter delta (Δ) with an extended base and its primary purpose was to serve as a steep staircase towards the heavens so that the Pharaoh could rejoin the Sun God Re; the extended line at the bottom of the sign probably expressed the desire for a broad base, but this idea was not developed: the pyramids would be embedded in the ground rather than broadened at the bottom. But a broad base was all the same a symbol of royal power: in the Cairo museum, where a series of everyday objects (gowns, sandals, etc.) used by Tutankhamen are displayed, we can note how the sandals he wore suddenly become broader from the moment the young Pharaoh (who died at eighteen) ascended the throne.

INTERNAL STRUCTURE OF THE SEVEN MAJOR PYRAMIDS

The originality of these pyramids lies in two of their features:

1) great load concentrations within a relatively narrow perimeter: a pyramid like that of Cheops weighs 5 000 000 tons and presses down on the soil within a square measuring 231 m along each side; in other words it weighs 6 500 tons per metre of its perimeter, far more than the heaviest structures of modern civilization – 300 to 900 tons/metre for a nuclear power station, 400 to 500 tons/metre for a building of 50 storeys;

2) the steepness of the slopes, inclined at about 52° from the horizontal plane and rising to a height of 60 m for Saqqarah and as much as 147 m for the Pyramid of Cheops.

Figure 17 shows the two-in-one slope of the downstream earthworks of a modern rockfill dam with the same height as the Pyramid of Cheops, illustrating the fact that present-day engineers are not nearly as bold as Imhotep and his successors were. Though such a comparison might seem unfair because a dam has to contend with the thrust of the water, this does not apply to the downstream filling.

The truth of the matter is that dams contain only heavily compacted raw materials (sand, stones and rocks) whereas the pyramids were built with a certain proportion of dressed stones and roughly squared blocks. In the case of the pyramids, the problem was to find the right proportion of shaped to rough materials that would ensure long-term stability, bearing in mind the size and skill of the workforce and the kind of stone available.

Once we depart from the ideal form of a parallelepiped and use blocks of an irregular shape, the latter will touch each other only here and there, inducing extremely high stresses at the points of contact. Some of these stresses are lateral and produce a sideways thrust which will tend to broaden and flatten the structure. It is true that the Egyptians were familiar with mortars made from clay and

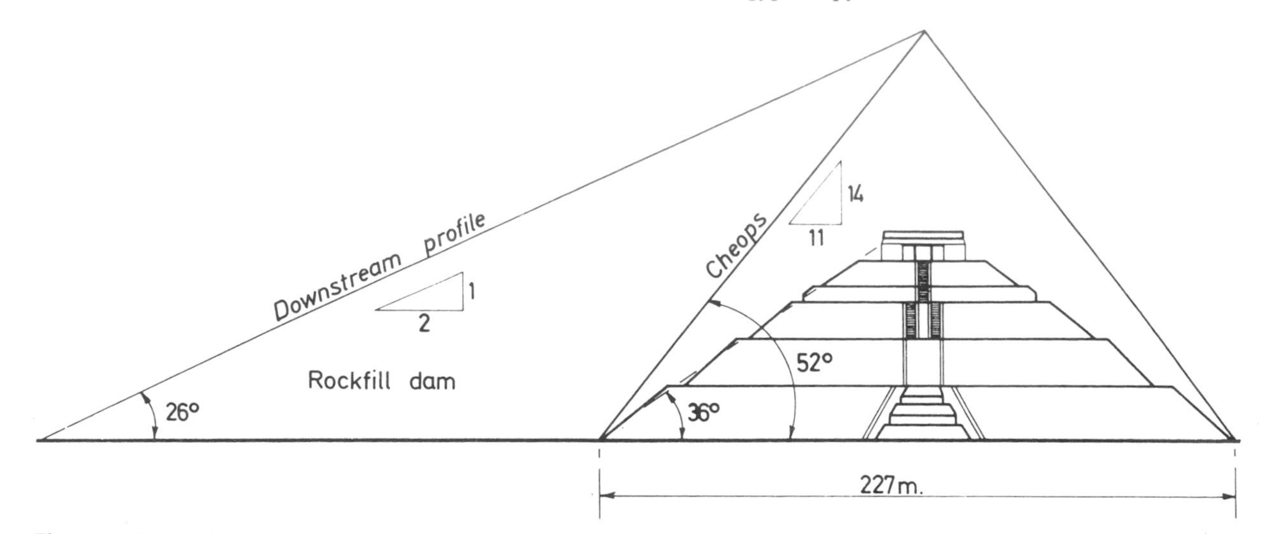

Figure 17. Comparison of slopes: the Pyramid of Cheops, the great Pyramid of the Sun in Mexico and a modern rockfill dam.

Photo 8. The Pyramid of Saqqarah with its six steps or stages (3rd dynasty).

powdered limestone mixed with gravel, sand and gypsum grains but these set so slowly that they could not be counted upon to annul the initial sideways thrust. Moreover, their strength was no more than a few percent of that of the stone.

The pyramid was built around a central nucleus or core the sides of which were inclined at 74° to the horizontal (Figure 19, left). This nucleus was composed of roughly quarried blocks surrounded by four masonry walls of dressed stone, numbered (1) in the diagram. Around this smooth surfaced core a first thick coating of masonry was added, again finished with smooth stone (2) at the same angle of 74°, and so on, with the summit of each successive coating of masonry lower than the one before, so that the final result was a stepped pyramid (Photo 8). Both at Saqqarah and Meidum, the prime numbers 2, 7 and 11 (Figure 18) were used to determine the slope of the walls and the alignment of the outer edges of the steps, which would later give the angle of the true pyramid.

In other words, the 'steps' are simply the tops of the external coating walls, which were designed to

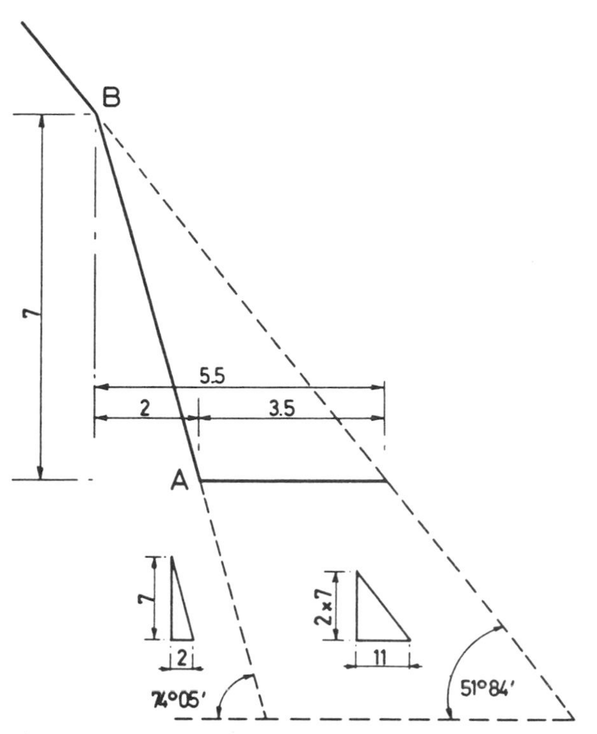

Figure 18. Profile of the step pyramids (Saqqarah and Meidum) showing how the slope of the coating walls and the angle at which the outer edges of the steps are aligned were based on the prime numbers 2, 7 and 11. The latter angle would later be used for the true pyramid.

Photo 9. The coating walls at Saqqarah (arrowed).

respect that satisfying proportion of dressed and roughly shaped stones. At Meidum, these walls are 5.2 m thick (10 cubits) with only about a fifth of that taken by the dressed stone. At Saqqarah, the coating of masonry still stands (Photo 9); the technique was invented by Imhotep but the blocks are small and fairly uniform in shape while the coating wall is only 2.60 m thick. There are many signs suggesting that the Egyptians were still not very experienced in the cutting of stone.

The reason for employing dressed stone for the outer surface of the coating walls – sometimes called buttress walls – does not appear to have been fully recognized. Its essential purpose – which explains the smooth external face – was to allow each coating of masonry to settle on its independent foundations without cracking the stones: however minute the differential settlement of the base under the weight of the heavy loads, which naturally decreased from the centre of the pyramid outwards, it was vital for these walls to be founded separately. Secondly, the masonry behind the dressed stone of each coating wall is more compressible on account of the more numerous and less regular joints, and thus tends to draw the wall downwards. A wall constructed of well dressed materials can transmit heavy pressure to the ground without risk of horizontal movement at the base owing to its narrow angle with the vertical plane (90°–74° = 16°). Lastly, the exposed upper surface of the walls (AB in Figure 18) has to serve much the same function as the walls of a silo, counteracting the thrust of the other materials used in the coating, at least until the mortar has set. It was in all probability this sideways thrust which determined the height AB and the inward-leaning angle of 74°.

In brief, the technique used for Aha's mastaba was taken further: the walls were now inclined at 74° towards the interior and no longer formed a network of square chambers but concentric squares at the base. Everything was thought out harmoniously: the geometry of the steps and the design of the coating walls took account of the possibilities of the workforce and their tools while ensuring good stability.

Meidum is a strange and enigmatic pyramid (Photo 10). Located off the beaten track it is one of the least visited by tourists; on the other hand it is one of the most fully explored by egyptologists. The shaded cross-section on the left of Figure 19 represents the pyramid E2, which was 85 m high

Photo 10. The Pyramid of Meidum in 1983, before the new excavations at its base. Seen from the northeast.

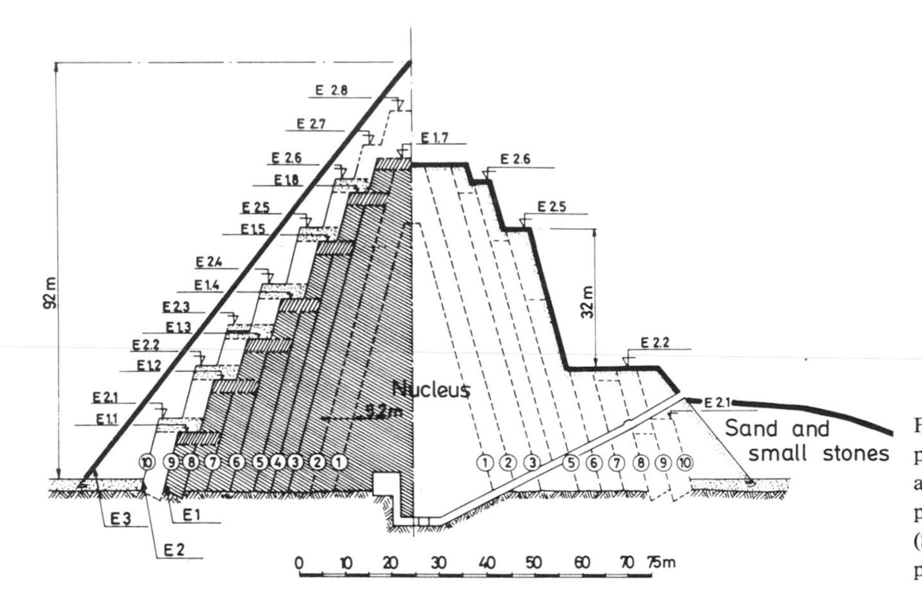

Figure 19. Cross-section of the pyramid at Meidum (Marogioglio and Rinaldi 1964). On the left, pyramids E1 and E2 (Huni) and E3 (Snefrou) and on the right, the pyramid in its present state.

and superimposed on the E1 pyramid. Both these pyramids were built by Huni. The various coating walls of masonry were identified by Wainwright when he dug a tunnel through the base of the pyramid from north to south. E3 represents the outer casing of the true pyramid built at a later stage by Snefrou. The right-hand side of Figure 19 is a cross-section of the pyramid in its present state: wall 6, built by Huni, is now exposed over a height of 32 m. In this wall an opening makes it possible to see the materials used within the coating (Photo 12), which are in this case roughly shaped and without mortar.

Recent excavations at Meidum have revealed the lower part of the E3 pyramid. Why then does the pyramid as a whole have this strange shape today?

Some blame it on pillaging and the theft of stone from the upper part, while Mendelssohn (1974) argues in favour of collapse as a result of rain.

The problem of Meidum, like many other problems related to the pyramids, has been studied mainly by teams composed for the most part of ar-

Photo 11. Northwest corner of Meidum after clearance of rubble in 1986. At the bottom and on the right may be seen the facing stones of the E3 pyramid.

Photo 12. The roughly quarried internal masonry blocks to be seen through an opening in wall 6 at Meidum (north face).

chaeologists, historians, geographers, philologists, etc., without the assistance of geotechnical engineers. However much we may regard it as a work of art, the fact remains that a pyramid is basically a pile of stones which depends for its stability – and for the stability of the ground which has to bear the weight – on the laws of soil and rock mechanics.

In examining the recently cleared base of the pyramid, the first thing that strikes the observer is that the materials used for its transformation into a true pyramid were of very poor quality, being excessively friable (Photo 13) and incapable of ensuring the long-term stability of the recess angle ABC of Figure 20. The quality of building stone is determined by the velocity with which it transmits vibrations. While the normal values are 5000 to 6000 m/sec for sound limestone, the measurements made by Toureng and Denis at the Laboratoire Central des Ponts et Chaussées de Paris, based on samples from E3 selected by the author, gave a velocity of only 2700 m/sec.

Moreover, an examination of the north-west corner of the collapsed pyramid reveals that the laying of the stones behind the outer casing (Photos 13

Photo 13. Friability of the stones used by Snefrou to fill the space between the E2 and E3 pyramids (northwest corner). Note the hard limestone of the E2 wall 10 at the lower right (arrowed). (Photo by author).

Photo 14. Step E21 (arrowed) at the top of E2 wall 10 (see Figure 19 left). Note the horizontal capping stone with its base inclined at 90°–74° = 16° (Photo by author).

and 14) leaves much to be desired: some joints are too thick and are filled with a mortar that, under the weight of the loads, could have exerted pressure on the E3 casing at the time it was added. Moreover, as is shown in Photo 13, a mixture of good quality and poor quality limestone from local quarries was used. The latter, laid between weak stones, has exfoliated and crumbled into dust.

Owing to the friability of the stone behind the E3 casing, local cracks appeared wherever it was not well bedded, to be followed by readjustment, further cracks and a progressive advance of the zone of weakness from C towards A, as indicated in Figure 20. In the upper part, the smooth surfaces of the E1 and E2 coating walls formed dangerous slip planes. This process could have been speeded up by earthquakes which, before the present era, were less rare in Egypt than is generally believed. Professor

N.N. Ambraseys, Professor of Engineering Seismology at Imperial College, London, has kindly provided a list of historical earthquakes in Egypt deduced from ancient commentaries or from the collapse of temples or colossi, and ending with the major earthquake in 365 A.D. which caused a tidal wave that drowned some 5000 people at Alexandria.

If the collapse had taken place at once, it is hard to imagine Cheops boldly deciding to build a pyramid more than twice as high. It is very probable that the pyramid remained intact for at least 1500 years since inscriptions made by two scribes of the 19th dynasty, which have been found near the burial chamber, make no mention of any accident to it.

As for the theft of the stone, why would robbers climb so high to steal stones of poor quality while leaving the dressed stone at the base untouched?

The collapse has also been blamed on rain but, even if Egypt's annual rainfall of 12 mm were concentrated in a single cloudburst, it is difficult to see how this could provoke internal stresses that would explain the accident. However, it is undeniable that this water drained along particular routes: the whitish stains above the opening in wall 6 suggest that the hole could have been caused by water.

In short, the most likely reason for Meidum's present state was a long-term collapse, which was mainly due to a poor choice of materials for the construction of the E3 pyramid. This long-term collapse was facilitated by the theft of the handsome limestone of the upper part followed by undercutting due to the erosion of the mortar and more friable limestone by sandstorms. The final collapse may well have been triggered by an earthquake.

Did Shakespeare glimpse this fatal moment when he had Macbeth cry out to the witches, as the hour of retribution approached,

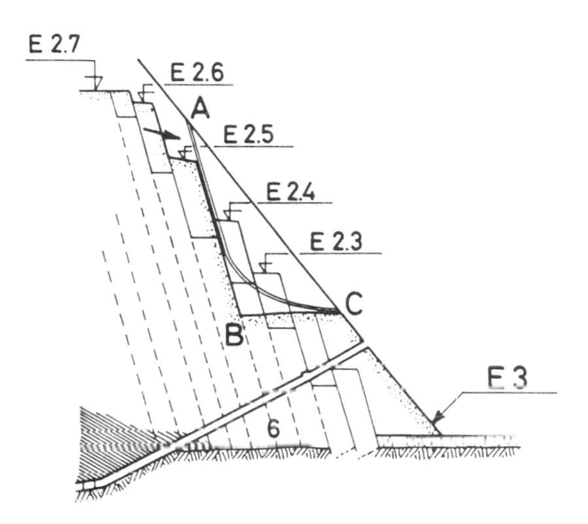

Figure 20. The double black line shows the possible slip line of the unstable wedge ABC of the Meidum Pyramid E3, vulnerable in the long-term.

Photo 15. Backing stones in fine limestone of the Pyramid of Cheops.

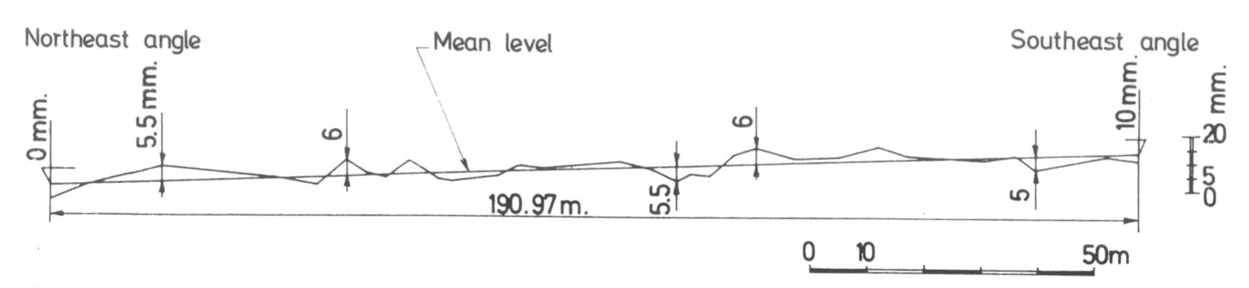

Figure 21. Pyramid of Cheops. Longitudinal profile of the second course of masonry, East face. Surveyed on 5 Feb. 1986 by Rondard.

'Though palaces and pyramids do slope
Their heads to their foundations...
Even till destruction sicken: answer me'.

Whereas the tunnel pierced by Wainwright has clearly revealed how Meidum was built, the structure of the three pyramids of Giza is still unknown. Their outer casing in splendid Tura limestone has been stolen, exposing backing stones that are even larger and more carefully shaped than the ones at Meidum (Photo 15). But are we to deduce that blocks of this quality continue right into the centre of the pyramids? The Pyramid of Cheops has been investigated many times but the studies have been mainly concerned with the precision of its siting and orientation and with the quality of the masonry in the backing stones. But what about the inside? Is there a nucleus? How are the differences in settlement – which the laws of elasticity require to be at their greatest at the centre – absorbed? As a matter of fact, to judge from that of the centre-points of each side of the pyramid, the settlement has been slight. At the author's request, Rondard, Chief Surveyor of the Greater Cairo Metro, checked the level at 35 points along the second course of stones on the eastern face. The line expressing the mean level of these 35 points (Figure 21) shows, firstly, that the difference in level between the N and S corners of the east face is only 10 mm; secondly, that there has been no appreciable settlement of the middle in relation to the two ends; and thirdly, that the irregularities in the laying (or cutting) of the stones for the second course do not exceed 5 mm up or down. Cheop's architect had clearly displayed great skill in both his choice of site and the actual construction. Perfection is a privilege of the rich – Mendelssohn claimed that Cheops had taken perfection so far as to give the stones a slightly concave alignment to prevent them from sliding outwards and thus improve the stability of his pyramid. (Figure 22.) As we see, this is not true.

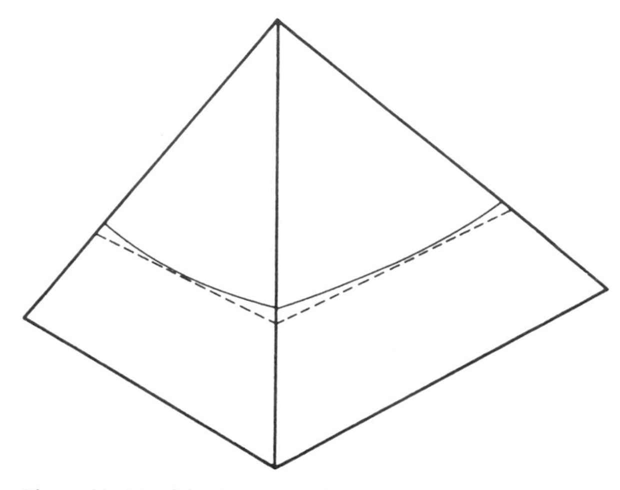

Figure 22. Mendelssohn assumed that the masonry courses of the Pyramid of Cheops were made slightly concave towards the sky. This is not true.

It has been estimated that the pyramid contains some 2300000 blocks weighing on average 2.5 tons each but with some as heavy as 15 tons. Clearly the cutting and shaping of all these blocks could not have been done without unshakeable metaphysical convictions on the part of the workforce or the coercive powers of a tyrant. This pyramid and the next one to be built – that of Chefren, son of Cheops – represented an enormous human effort which sparked comments that smack of jealousy and defamation.

Pliny the Elder speaks of 'regnum pecuniae otiosa ac stulta ostentatio' – 'the Pharaoh's vain and stupid show of wealth'. Herodotus is more perfidious: Cheops is said to have been extremely dissolute, with methods of financing his enterprises that could hardly be called moral. 'The depravation of Cheops reached such a point that, having squandered all his treasures and wishing for others, he sent his daughter into a brothel with instructions to bring back a certain sum of money, the amount of which I do not know because it is not mentioned. Nevertheless she raised the money and at the same time, resolved on leaving a memorial to herself, demanded a stone block from each of the men, with which she built the middle pyramid of the three situated in front of the large one. The sides of the base of her pyramid measure 150 feet each'. All that needs to be said is that this pyramid contains 20000 stones!

Despite the attacks and the methods used, the great undertaking of Cheops, like the other two pyramids of Giza, stands proudly in the desert, guarding the mystery of its internal structure. Why should Cheops reject the previous techniques? Borchardt (1911) could see no reason for him to do so but Clarke and Engelbach (1930) take the opposite view – and discussion continues. Some very recent measurements by microgravimeter inside the pyramid (Delétie) have revealed an overall average density of about 2.0, much lower than that of the backing stones. Did Cheops follow the Imhotep technique of internal walls, but with very rough blocks laid between them? If so, the long-term stability of the pyramid could have been aided by the use of a quick-setting gypsum paste mortar like that found in the King's Chamber (Regourd).

INTERNAL STRUCTURE OF THE LATER PYRAMIDS

The pyramid of Mycerinus is only one ninth of the volume of Cheops and all the following pyramids were not only smaller but of inferior construction. Neferirkarè, the largest of the three pyramids built at Abu Sir during the 5th dynasty, was 68.4 m high and 108 m square at the base. Very recent clearing of rubble has revealed the existence of internal walls (Photo 16). Do these extend down to the bottom as at Meidum or are they shorter retaining walls simply staggered back on top of one another before the transformation into a true pyramid? It is impossible to say without further excavation. Another pyramid of this group of three, that of Sahurè, has been explored by Borchardt and Petrie; the former regarded it (Figure 23) as a faithful expression of the ideas of Imhotep.

The pyramid of Unas (last Pharaoh of the 5th dynasty) reveals a wall in the Imhotep style descending to the base.

By this time, however, the Pharaohs had long lost part of their power and wealth and the mobilization of the peasants to build such pyramids had impoverished the country. It is thus hardly surprising that Pepi II (6th dynasty) should greatly simplify his own pyramid at Saqqarah; it was built of small stones jointed with a gypsum mortar and faced with heavy Tura slabs toed into the rock to contain the lateral thrust of the filling. As the Tura stone has been stolen, the pyramid has naturally suffered from the weather and is now just a heap of sand and small stones.

It was not until the 12th dynasty and Sesostris I that some new ideas were tried out. His pyramid at Lisht was 60 m high and 105.6 m square at the base. Eight walls of dressed stone which radiate out from the centre to the four corners and the mid-points of each face form the basic structure. These

Photo 16. Pyramid of Neferirkarè (5th dynasty) at Abu Sir. Very recent clearing of rubble has exposed the internal walls.

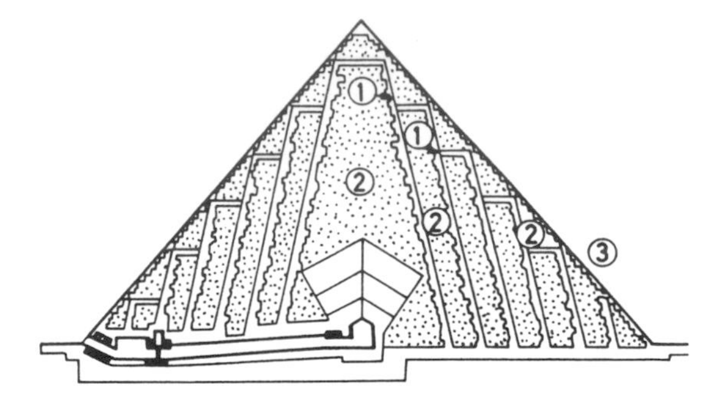

Figure 23. The Pyramid of Sahurè (Abu Sir, 5th dynasty) in its original state (after Borchardt).
1. Internal facing stones of fine limestone
2. Rubble filling
3. External facing of polished Tura limestone.

walls delimited eight compartments, each of which was then subdivided into two unequal parts by secondary walls equidistant from the corners and the mid-points. The sixteen compartments were then filled with rubble and the whole covered with a heavy casing of fine-grained limestone. This approach represents an adaptation of Imhotep's method that is valid for a pyramid of moderate size. At El Lahun, at the edge of El Faiyum, Sesostris II took this idea further: his pyramid contains walls laid out like a grid and is anchored in the rock, this time in order to contain the thrust of sundried bricks. Shortly afterwards the Pharaohs, discouraged by thieving from the burial chambers, opted for underground tombs cut into the rock of the Valley of the Kings. But the ideas developed by Sesostris I and Sesostris II were to live on for a long time in the pyramids built in the extreme south of Egypt towards the Sudan. These pyramids are very pointed, with their lack of height compensated by the vigour of their upward reach (68° to the horizontal). The external facing was in limestone and later, at Meroe (after 200 A.D.), in bricks covered with plaster. The inside was divided into compartments like silos, which were filled with stones.

In the history of the pyramids, the Pyramid of Cheops is exceptional in size, in the materials used, and in the bonding of the stonework – all of which required an immense workforce. The subsequent decline reflected that of the Pharaoh's power and the emergence of social and economic needs which urged a return to the Imhotep approach, but with the use of inferior materials; finally, apart from the heights and the materials employed, it was back to square one – the mastabas of the 1st dynasty and King Aha.

If there is one type of monument in which the 'Down to Earth' problems are interesting to examine, it is surely the pyramid. That is why we have spent so much time on the subject. Alain wrote: 'The pyramid was constructed concentrically by adding a casing to the basic form, always keeping to the same shape which is that of a heap of stones in equilibrium.' This is a superficial view of the matter, since the shape of a heap of stones in equilibrium depends entirely upon the effort made to chisel those stones from crude lumps into flat-sided blocks. In the free-standing mass of the Egyptian pyramid the intelligence of the internal structure is concealed, and we are left to admire the subtle compromise between the upward surge towards the heavens and the human toil to give the stones a form.

THE BURIAL CHAMBERS

The pyramid is the housing for a burial chamber, which is the final purpose of the monument. But whereas the architecture of the actual pyramid pursued a logical development during the period with which we are concerned, the positioning and size of the burial chambers suggest greater uncertainty as to the right approach.

Figure 24 shows where they are sited: whereas the small (1.65 m wide) chamber of Zoser is cut into the rock 28 m below the surface, his successors wanted to find refuge after death at the base or within the pyramid itself. Were they aware that it is much more difficult to place a tomb with rigid walls within a relatively more compressible structure? Such a cavity induces stresses against its ceiling and walls which are absent when it is located within a mass of rock. This concentration of forces becomes more and more dangerous as the burial chamber gets larger and the weight of the pyramid increases (Figures 24 and 25). Care must therefore be taken not to cover the chamber with a flat slab of stone, even one of Aswan granite, and this the Pharaohs realized. Their burial chambers would have corbelled roofs (Huni, Snefrou, Cheops), made by stepping a few inches inwards each of the upper courses of the walls leaving at the top a very small span to be covered; then, to simplify matters, Cheops and Chefren added a pointed roof formed

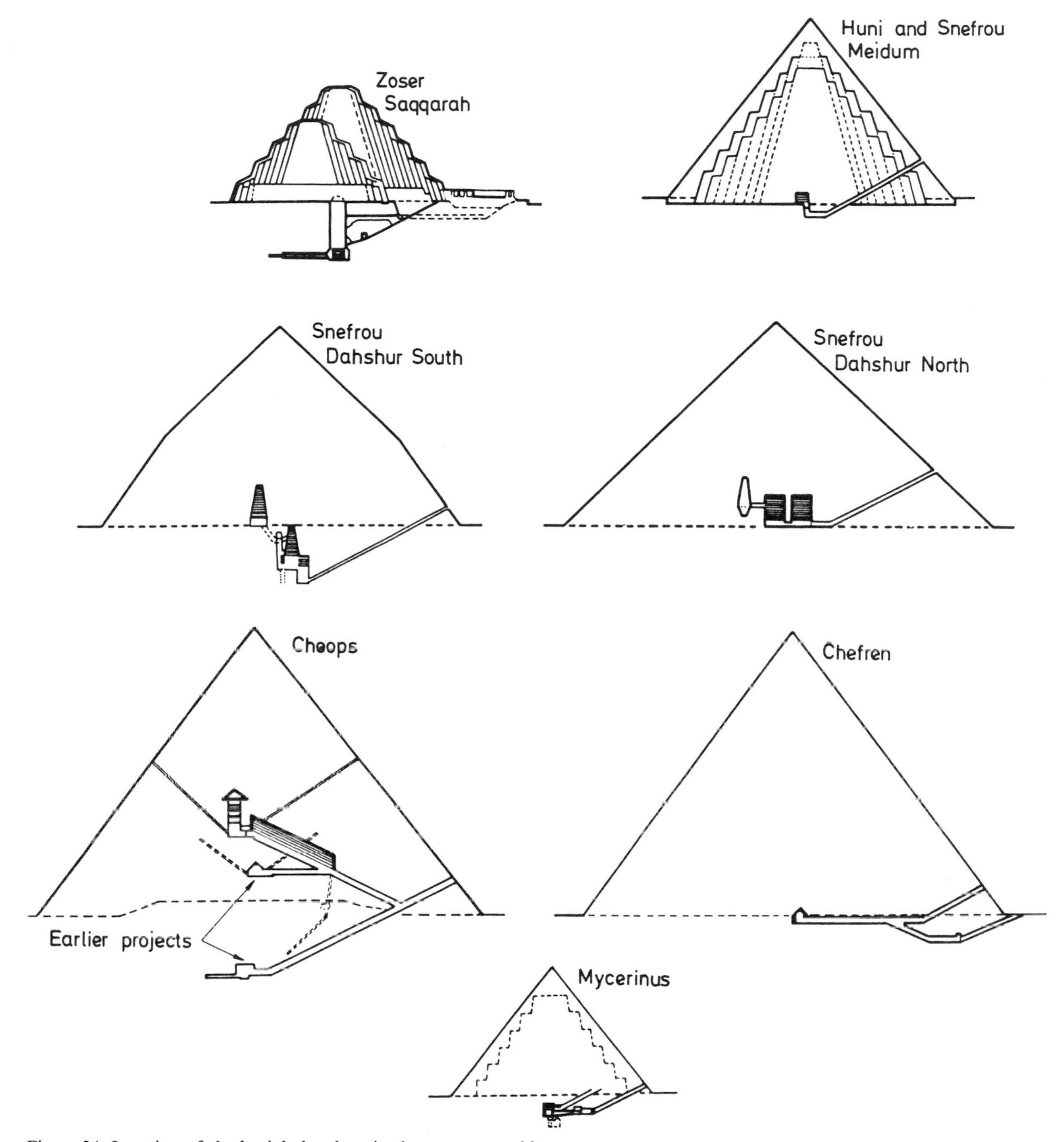

Figure 24. Location of the burial chambers in the seven pyramids.

of slabs leaning against each other.

However, this was not enough: if the full height of the chamber and corbelling is set within a nucleus of rough masonry that is relatively more compressible, the masonry will exert lateral pressure on the chamber. This explains what happened at Dahshur South, a pyramid we shall now consider since its shape was dictated by the burial chamber inside.

A Pharaoh usually built a single pyramid – for his tomb. Snefrou, however, founder of the 4th dynasty, not only built two at Dashur, the Bent Pyramid to the south and then the first tetrahedral pyramid to the north, but he also finished the one at Meidum.

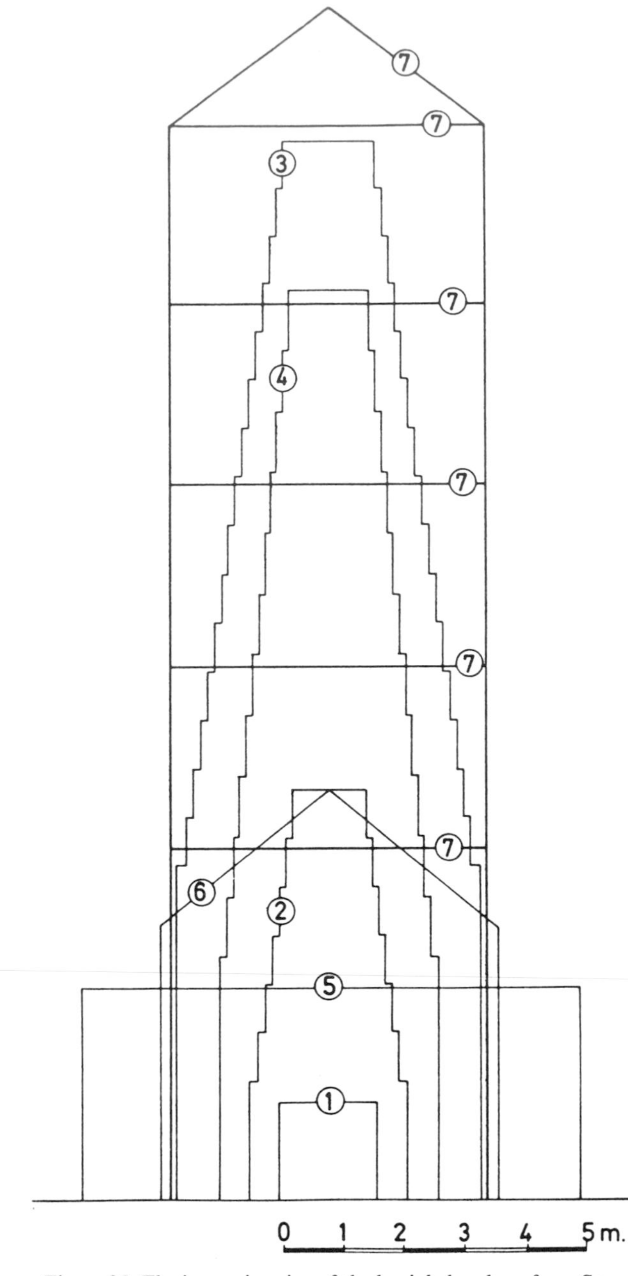

Figure 25. The increasing size of the burial chambers from Saqqarah to Cheops.
1. Saqqarah; 2. Meidum; 3. Dashur South (upper lower chamber); 4. Dashur North; 5. Cheops (subterranean chamber, after Vyse and Perring); 6. Queen's chamber; 7. King's chamber.

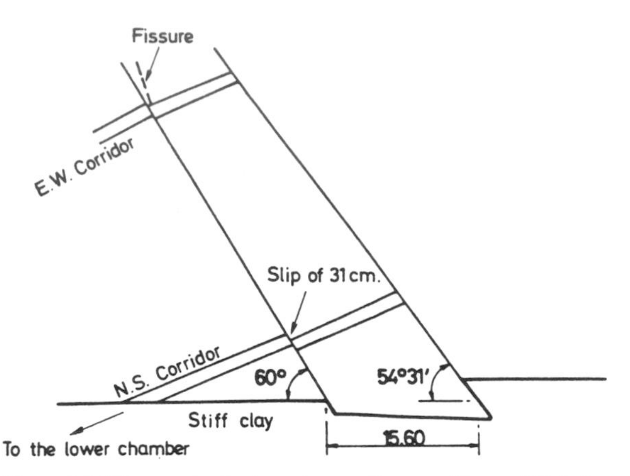

Figure 26. Slippage in the two descending corridors of Dahshur South.

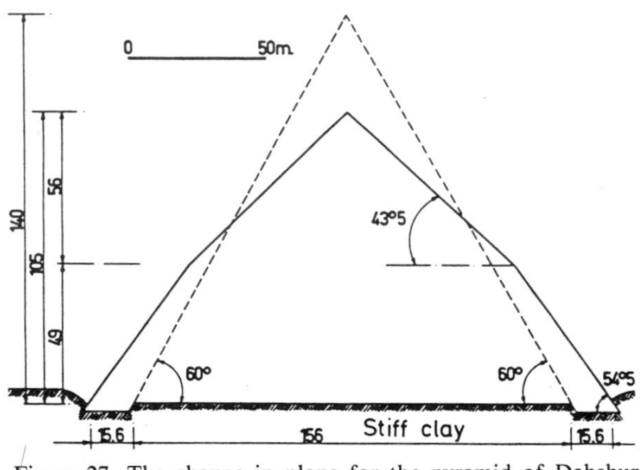

Figure 27. The change in plans for the pyramid of Dahshur South.
- - - - the true pyramid as originally planned

For the South Dahshur pyramid, Snefrou first tried to construct a true pyramid with sides inclined at 60° to the horizontal (Figure 27). This very steep Pyramid therefore had a severe punching effect on the soil but, in addition, it was built on an unsuitable site, on clay that was doubtless stiff but deformable under a heavy load. Uneven settlement took place, causing fractures and slip that can be seen today (Figure 26) in the two NS and WE descending corridors each leading to a burial chamber.

Both of these chambers were 4.95 m wide (nearly twice the width at Meidum) and covered like Meidum with a corbelled roof. The deeply buried chamber, unlike the one inside the pyramid, has stood up well to the centuries. The relatively poor quality of the substratum, the disturbance of the soil to build the first chamber and the location of the second chamber in the core led to lateral pressure on the latter's walls. Snefrou had to fit stout cedarwood poles and build up the floor to stop the walls from caving in.

For anyone who has worked underground, the cracking sound from wooden props and the sharp report as a rock breaks have an ominous note. Now Snefrou, as Wildung (1969) remarks, was a 'bon vivant' and had gained the reputation of being a good king. The translation of the Westcar papyrus enabled Derchain (1969), in an article on 'Snefrou and his oarswomen', to describe how the Pharaoh used to enjoy having his royal barge propelled along the Nile by 'the sexiest members of his harem clad only in a hairnet!'

It is not surprising that Snefrou, who loved life and beauty, should decide, when seeking the ideal solution for a true pyramid, not to use these burial chambers but, before constructing a new smooth-sided pyramid, to finish this one by making certain adjustments to the design: easing the slope to 54°31 (Figure 27) and then, from a height of 49 m, a further reduction to 43°5, which explains the 'bent' form.

Oddly enough and in spite of these incidents, the pyramid, which nonetheless attains a height of 105 m, seems to be one of the best preserved, with its external casing of Tura limestone spared by thieves, unlike the case nearly everywhere else. The stones used were also bigger and better set than those of the 3rd dynasty. Closer inspection, however, reveals that all the blocks contain numerous fissures crossing a good many courses of stonework.

Snefrou designed his second pyramid, that of Dahshur North, with a single slope at the same angle as the upper part of the Bent Pyramid (43°5) and reduced to 4 m the width of the larger of the two burial chambers, which were covered with a corbelled roof. The site was no doubt firmer and it is highly probable that this time Snefrou took the precaution of surrounding the burial chambers, at the time of construction, with carefully dressed stone to provide an effective lateral support. He was so successful that these burial chambers are still in a wonderful state of preservation.

Faced with these very different results Cheops hesitated. He first planned a chamber (Figure 24) 30 m underground, but he left it unfinished. His second idea was to build a chamber, now called the Queen's Chamber, inside the pyramid and 30 m above ground level. With a span of 5.24 m, it was the widest so far constructed and was covered by a pointed roof; although it proved satisfactory, Cheops opted for a third solution by building a burial chamber even higher up the pyramid. Constructed of Aswan granite, this chamber is 5.15 m wide and 5.80 m high. The form of the fissures in the walls suggests that the chamber is supporting very heavy loads. The roof is made of granite slabs which are now cracked. Above these slabs are four relieving compartments, each of which is topped by granite slabs – all now cracked – and a fifth compartment with a pointed roof.

Why all these cracks? Are they the sign of a series of unsuccessful attempts to cover the chamber with a flat roof or of the fact that the chamber is located very high up within a deformable core? It is difficult to be sure without exploring laterally. Whatever the answer, one is struck by the contrast between these cracks and the almost perfect condition of the famous corridor leading up to the King's Chamber; it is exceptionally wide (7.20 m) with walls rising vertically for 2.25 m surmounted by a corbel vault with seven successive inward-projecting courses, the last and most majestic example of this technique. This corridor is certainly not located in the nucleus but is probably set in a thick shell of precisely adjusted masonry which protects it from stresses.

Taken as a whole, the general approach of Cheops appears undecided and imbued with a sense of grandeur. After him, Chefren would have

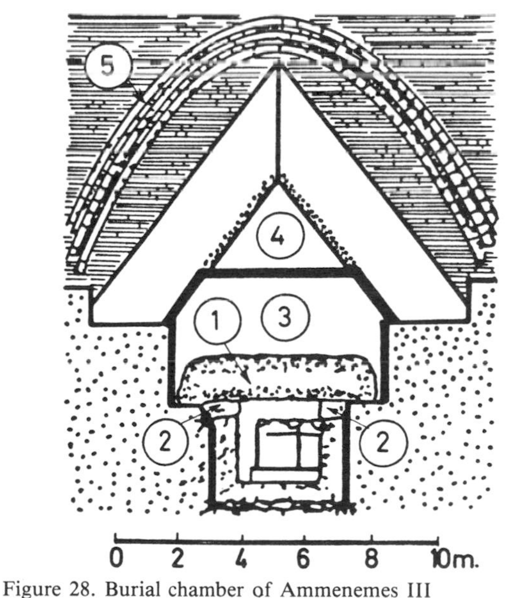

Figure 28. Burial chamber of Ammenemes III

1. Three slabs of yellow quartzite
2. Blocks at the top of the walls to give extra height
3. Lower relieving compartment
4. Upper relieving compartment with a pointed roof
5. Brick arch 90 cm thick.

(after W.M.F. Petrie).

the walls of his burial chamber in the ground, with only its pointed roof within the actual pyramid. All subsequent burial chambers, with only the odd exception, would be underground: back to Imhotep's prudent solution for Zoser.

The most well-known exception is that of Sahurè (Figure 23). Here the burial chamber is once again set in the middle of the pyramid and covered with a pointed roof: Borchardt discovered three superposed slabs on each side and Petrie estimated that the largest of these slabs is 10.50 m long, 2.70 m wide and 3.60 m thick. Only two of them are still intact.

From the 6th dynasty onwards the burial chamber was once again located in the rock, and with Sesostris III and Ammenemes III (12th dynasty) at Dahshur, the use of kiln-fired bricks for the nucleus made it possible to transform the centre into a labyrinth of corridors leading to underground chambers. This approach was taken further in Ammenemes III's second pyramid at Hawara. The Pharaoh was a past master in the art of throwing robbers off the scent: the pyramid has false doors, corridors leading nowhere and stone-filled shafts with no exit.

The actual burial chamber is truly remarkable (Figure 28). Before a start was made on the superstructure, a vast rectangular pit was chiselled out of the rock and then lined with stone. A huge monolithic quartzite sarcophagus without a lid (internal measurements: 6.60 m long, 2.40 m wide and 1.10 m high) was then lowered into the space. The roof was composed of 3 quartzite slabs 1.20 m thick laid side by side on the rim of the pit with stone wedges at the head of the chamber. Above this were two relieving compartments, the upper one of which had a pointed roof formed of two fifty-ton slabs of limestone. Lastly, this was topped by an enormous parabolic arch of brick 90 cm thick.

This remarkable construction – a little overdone perhaps – directs the downward thrust of the pyramid towards the ground well away from the burial chamber and foreshadows the future development of arches.

* 4 *

The art of foundations: Greek elegance and Roman solidity

THE GREEKS

Under the brilliant sunlight of the Mediterranean climate man prized the shade and the portico became the central feature of Greek architecture (Photo 17): public buildings were surrounded by stoas with their rows of slender and evenly spaced columns with capitals to bear the weight of the architrave, the frieze and the cornice (Figure 29).

In the development of Greek architecture, we generally distinguish the squat and rugged Doric from the light and elegant Ionic. But whereas these two orders strove in turn to win the favour of the Greeks, the loads concentrated on the columns were always spread over the ground in what may be called a Doric approach, with sturdy and efficient foundations.

Each column rested on the orthostats (Figure 29), which are long blocks of dressed stone laid on their sides in two or three layers to form the upper part of a foundation that broadened out at the base. This allowed the concentrated loads in the columns to be transmitted to the soil with a low and even pressure. In this Doric order, the base of the column corresponds to the capital, the orthostats to the architrave and frieze, and the deeper foundations to the cornice.

The orthostats were attached to each other with iron cramps (Figure 30) partly in order to spread the load more evenly but also to prevent the foundations from being dislocated by earthquakes. These iron cramps grew out of the Greeks' meticulous observation of such natural hazards. Indeed, it would be fair to say that they raised the design of earthquake-resistant architecture into an art.

Greek literature, particularly the prose of Eratosthenes, contains many detailed descriptions of earthquakes and the damage they caused, but the highly observant turn of mind characteristic of the Greeks enabled them to invent some most ingenious ways of countering their effects.

At first they tied the outside walls either by dovetailing the stones or by embedding wooden tie-beams in the masonry.

Later, when they had learned how to shape copper and iron, they reinforced the joints with the cramps mentioned above. These cramps are found not only in the foundations but also high up in the architraves and cornices, with the joints staggered if more than one layer was laid.

But the history of these cramp-irons does not stop there: in the restorations of the Acropolis monuments, especially those undertaken by Balanos between 1898 and 1933, similar iron cramps were used to reinforce portions of the architraves, etc., which were tending to break up. However, the restorers did not pay enough attention to the care with which the ancient Greeks had embedded their cramps: the hollow which the latter cut in the stone to receive the head of the cramps was made slightly oversize in order that a film of molten lead could be poured in so as to completely sheathe the iron, protecting it from rust and at the same time protecting the original stone by absorbing any tensile stress at rest (due to an error of assembly for instance) so as to reduce the effect of an earthquake. The cramps used by the early twentieth-century restorer were much less carefully made: they became rusty by the 1950's, which caused them to swell and split the marble. They will shortly be replaced by cramps made of titanium, which does not oxydize even in sea air and which has a low coefficient of thermal expansion approaching that of the marble.

In the case of opus incertum masonry, the facing was composed of irregularly shaped interlocked stones in order to avoid planes of weakness (Figure 32).

Photo 17. Reconstruction of the Attilus stoa in Athens (permission by the American School of Athens).

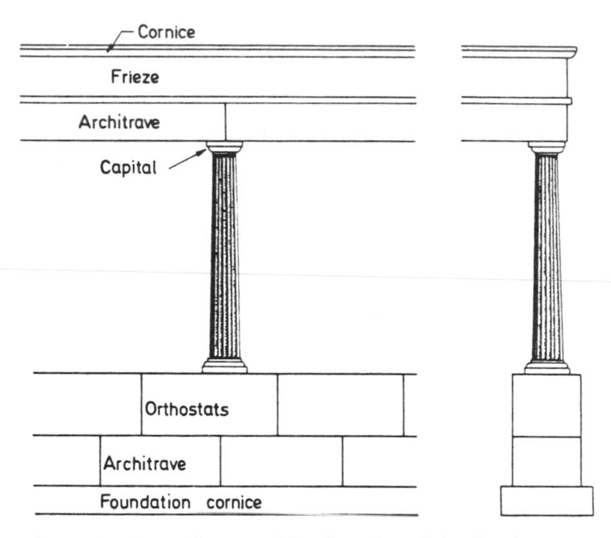

Figure 29. The underground Doric order of the Greeks.

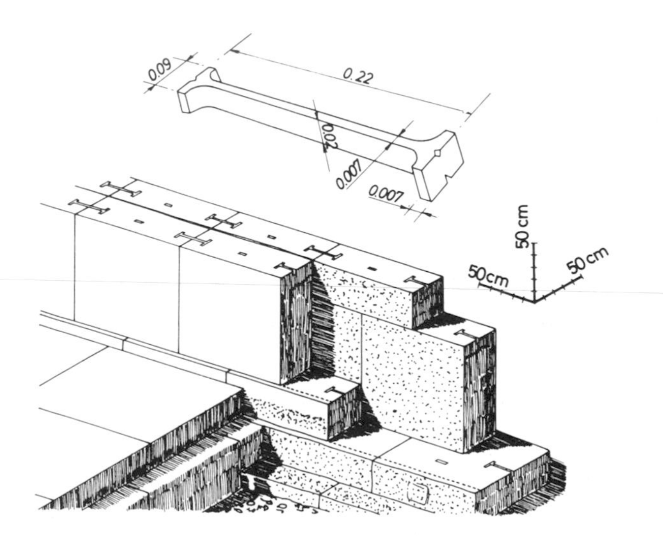

Figure 30. Detail of the orthostats at Delos.

Lastly, the Greeks were already aware that clays and soft substances damp vibrations. For example, the Temple of Diana at Ephesus, the Artemision, rebuilt in 560 B.C. after a fire and considered by Pliny (XXXVI–95) as one of the wonders of the world was, he relates, erected 'on a marshy site so as not to feel the effects of earthquakes or be vulnerable to crevices in the ground'. Moreover, as the Greeks did not want the foundations of such a massive structure to rest on a slippery and unstable soil, they first spread a layer of compacted charcoal over the soil, which they then covered with fleeces.

It was thus the Greeks who invented the first vibration-absorbing pads. As we shall see later, they also created the first geophones.

After the twentieth century invention of reinforced concrete, the Greek approach would be revived in the form of a strip footing (Figure 31),

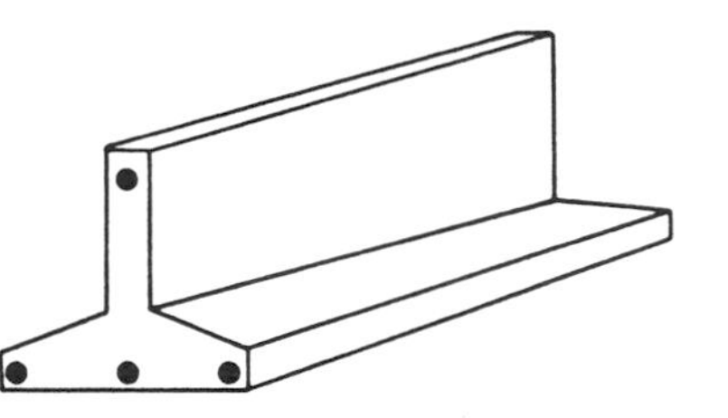

Figure 31. A modern equivalent to the underground Doric order of the Greeks: the strip footing in reinforced concrete of the twentieth century.

Figure 32. Use of curved polygonal blocks (underground tank in the Acropolis). 6th Century B.C. From R. Martin, Manuel d'Architecture Grecque, Picard, Paris, 1965.

which is nowadays a very common type of foundation that ensures the effective distribution of much greater concentrated loads.

Where the underlying soil was of poor quality, rendering this method of approach inadequate, the Greeks did not hesitate to clear away the top spongy layers until they reached a suitable base. On this they would lay several layers of large dressed slabs the joints between which were reinforced with cramps and staggered from layer to layer in accordance with special rules (Ginouvès 1966).

The modern world has adopted and developed this arrangement by laying a thick monolithic base in reinforced concrete (called a raft) which evenly distributes the considerable loads represented by blast furnaces, nuclear power stations and so forth.

THE ROMANS

The Romans added nothing to man's knowledge of the materials forming the earth's crust: like the Greeks, they attributed to stones the power to transform and reproduce themselves – the hardest ones being regarded as male and the soft clays as female. We have to wait until Linnaeus and the eighteenth century before such beliefs were finally scotched: 'Lapides... corpora congesta nec viva, nec sententia' (Rocks are solid bodies, they are not living and they have no senses). But, like the Greeks and in spite of such beliefs, the Romans were good architects – though less inspired – and considered it important that both the visible and non-visible aspects of what they built should perdure. Their central concern was 'firmitas' – solidity – the idea on which the teaching of Vitruvius in his 'De Re Architectura' is based.

This 'solidity' was imposed by earlier setbacks. During the first centuries of their capital's growth the Romans, for reasons of economy, had fallen into the habit of relying on the sun-dried bricks so widely used in the preceding millennia. But the density and height of the buildings erected within the walls of Imperial Rome increased rapidly and when, for instance, the Tiber flooded in 54 B.C. the brick foundations of many of them were washed away and great damage caused. Indeed, if we are to believe Seneca (4 B.C.– 65 A.D.), who remarked that the collapse of one's house was a common misfortune that could befall anyone during his lifetime, it was by no means rare for buildings to fall down. Eventually, an edict was issued prohibiting the use of raw bricks and so the Romans developed the use of good quality stone, which they employed with a variety of techniques for their walls and powerful vaults. At the same time they made greater use of kiln-fired clay in the form of tiles one foot or two thirds of a foot long (pedali and sesquipedali) and invented concrete.

The word 'concrete' comes from the Latin 'con-

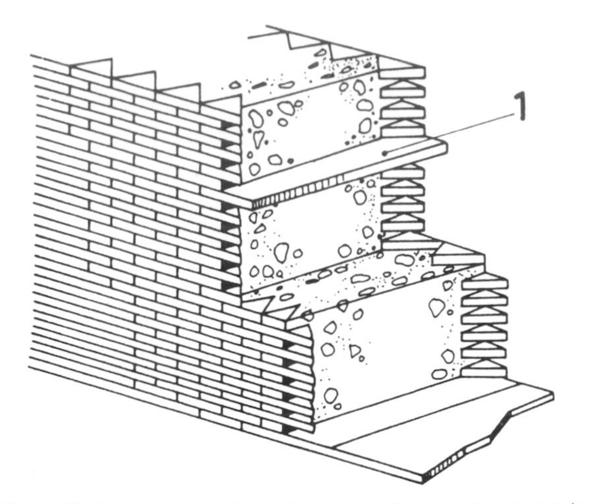

Figure 33. Roman concrete cast between a formwork in brick for foundations. 1. Wooden tie-bar to counter the thrust of the concrete

crescere', which means 'to grow together'. The Romans shaped the blocks of stone quarried from the hills overlooking Rome and stumbled on the idea of mixing the quarry chippings with the pozzolanic sands of the region and lime manufactured by the heating of limestone. Though favoured by an abundance of materials in their exceptional geographical situation, the Romans were also inventive. This concrete, however, set very slowly and ways had to be found of containing its thrust after it had been cast (Figure 33).

One outstanding success with its use was the laying of the concrete raft for the Colosseum. This was a circular slab of concrete 170 m in diameter sited on the Domus Aurea Neromana, a former lake which had been dug in order to collect the run-off from three nearby hills, the Celio, the Esquilino and the Palatino. The Cloaca Maxima sewer, located 11 m lower and only 500 m away, made it possible to drain the lake and cast the concrete on sound rock, a lithoid tuff.

The Romans were great builders who also used iron cramps to reinforce their structures, though to a lesser extent than the Greeks, but it never occurred to them to strengthen the concrete with iron, no more than it did to the generations in the seventeen centuries that followed. As a result, they sometimes encountered difficulties arising from their concrete's lack of tensile strength, even in foundations. This fact simply increases our admiration for their boldness in using non-reinforced concrete for the great dome of their Pantheon.

However, it was in road-building that the Roman civilization completely outclassed its predecessors.

Roads have a very long history. Excavating in the wetlands of Somerset, B. and T. Coles (1986) have recently discovered one of the world's oldest roads, the Neolithic Sweet Track. The Egyptians built a causeway leading to the pyramid of Unas, paved with stone slabs, which still exist, and we have already spoken of the Procession Way in Babylon, where Nebuchadrezzar II, boasting of his conquest of Lebanon, wrote: 'I have cut through steep mountains, I have split the rocks, I have made a way through and built straight roads for (transporting) the cedars'.

For the Greeks in particular, the sacred ways were somewhat rudimentary, being constructed of stone slabs laid out in two parallel lines, with grooves cut into them at a spacing of between 1.38 m and 1.44 m to guide the ox-drawn carriages (Figure 34).

The Romans made rapid progress in this aspect of engineering, eventually leaving us a remarkable network of about 90 000 km of primary roads criss-crossing their Empire, together with some 200 000 km of secondary roads (viae vicinales). Many of the highways have survived to the present day, forming one of the most familiar vestiges of the ancient world.

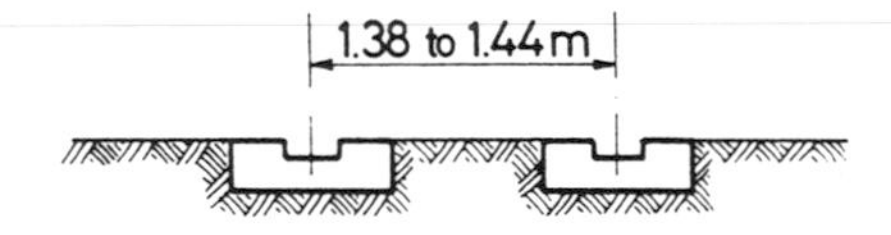

Figure 34. Greek road with two grooves cut into the stone to guide the carriages; the ancestor of present-day railways.

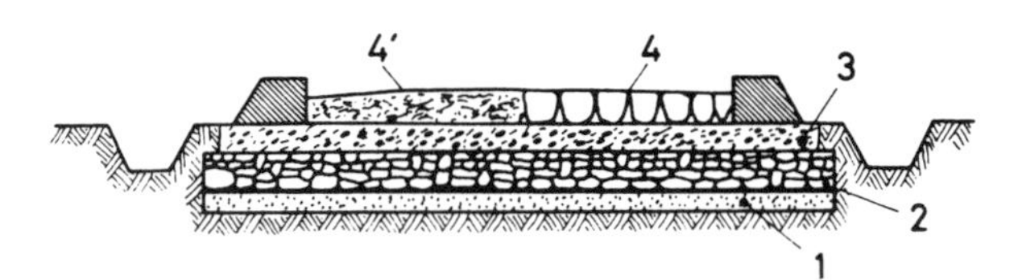

Figure 35. The Roman military road: a system of construction still unsurpassed.

1. The 'statumen' (20 to 30 cm thick): a layer of mortar over a layer of sand.
2. The 'rudus' (30 to 50 cm): slabs and blocks of stone with cement mortar joints.
3. The 'nucleus' (30 to 50 cm): gravel and broken stones mixed with lime to form a kind of concrete.
4. The 'summum dorsum': either stone slabs (4) or gravel concrete (4').

There are three points of interest in the construction of the great Roman roads: firstly (Figure 35), the basic design in the form of four superposed layers; secondly, the fine sand used in the first layer (to prevent underlying clay from rising) and the hardness and tightness of the fourth layer (to render it impervious to wear by rain and wheels); thirdly, the fact that they were designed in a rational manner to cope with the traffic expected. With the industrial revolution and now the age of motorways, the Roman approach still underlies the design of highways, with but a few minor changes with respect to the materials used.

In the meantime, however, the world's road-builders forgot the lessons taught by Rome and thought they could dispense with the bottom layer (the protective layer which prevented the clay from rising) and the third layer (the firm nucleus), with the result that their roads soon deteriorated into a string of potholes.

* 5 *

A refined approach to foundations: China in the first millennium

In many countries, the first millennium of our era was a period when the lessons of the past sank into oblivion and were replaced by the belief in myths which often bore little relation to the real problems of builders. In India, a book on architecture, written in Sanskrit by an unknown author, probably of the seventh century, and translated into English in 1964 under the title 'Architecture of Manasara Shilpashastra', gives many of the numerous myths and customs related to the founding and construction of buildings.

The first act of the architect was to drive away the evil spirits which haunted the site: 'Let all creatures, demons and gods as well, leave this place; let them go elsewhere and make their abode there'. This is followed by a large number of rules, such as:

'The wise architect should wash the excavations with the five products of the cow'.

'The architect should put on his best clothes and outer garment, and worship the Lord of the Universe with perfumes and flowers and then meditate on Him'.

'The excavation should be made at night and the bricks should be laid in the daytime'.

'The chief architect should fix the male bricks in the temples of male deities'.

Some ethnologists maintain that, in the East, the aim was in most cases to assemble the magic forces of the site and to transform them into allies of the human dwelling so that the building would be in harmony with the cosmos.

Curiously, however, this myth-dominated millennium engendered certain constructions in the Far East which testify to great skill and creative genius. China is a case in point. This country has recently developed a vigorous interest in archaeology and, what is more, does not lack boldness: the China Civil Engineering Society has not hesitated to conduct investigations in the immediate vicinity of some remarkable ancient monuments, a risk we would scarcely dare to run on account of their extreme vulnerability.

This has enabled the Chinese to throw some light on the techniques employed during the first millennium. The experience gained in the course of that period of great activity was codified in 1103, during the Sung Dynasty. This code, which contains some very interesting details on the art of building in a country subject to earthquakes and powerful storms, runs to no less than 3555 articles spread over 34 chapters. One of them is concerned with the damming of the soil (what we now call compaction) and Figure 36 illustrates the variety of techniques used at the time for this purpose. Admittedly, this was not a new idea – it is also to be found in Vitruvius' 'De Re Architectura'; in Book III, Chapter 3, for example, Vitruvius states 'If the soil is not firm, it should be made firm by compacting it, employing a machine for installing piles' and in Book VIII, Chapter 7, he advises the use of wooden piles reinforced with iron.

The originality of the Chinese, however, lay in their success in adapting this approach to all soils, including fat clayey soils which, when compacted, react rather like a soft pillow. The Chinese first rendered them firmer by alternating soil with stones and broken bricks: the hole dug out for the foundations was filled with a layer of broken stones, a layer of clayey soil, a layer of broken bricks, and so forth, with each layer carefully compacted. They even went so far as to specify the proportions by weight of these layers and by exactly how much proper compaction should reduce their volume. The effect of the compacting was to fragment the bricks so that they would eventually cement together somewhat to form a solid base.

It was a technique of this sort which, between 605 and 617 A.D. and hence long before the Sung

Figure 36. The range of methods used to compact the soil, described in the Sung Code (1103).

Code, probably permitted the construction of the magnificent Anchi bridge (Photo 18), to which the author's attention was drawn by Professor Lu Zhao-Jun. In purity of line it more than equals the most beautiful stone bridges of the nineteenth century. As one can see, voids have been formed by secondary arches in the spandrels in order to reduce the dead weight on the flat arch and allow the passage of water at times of flood.

The result is a pleasure to the eye, but the non-visible architecture is just as remarkable for its intelligence. It is well known that the flatter an arch (Figure 37), the greater and the more horizontal is the outward thrust exerted on the ground by its abutments. Clearly, any outward movement of the abutments rapidly causes irreparable damage to a flat arch since the keystone quickly gives way.

The ground on which the piers of this bridge were founded is a rather fat – hence deformable – clayey soil. For it to survive 1300 years of earthquakes, floods and growing traffic, Li Chun, the architect of this masterpiece, had to treat and compact the soil all around and under the piers in the manner illustrated, an audacious plan which has met with complete success.

Another monument which has recently attracted the attention of Chinese archaeologists is the Pagoda of Longhua (Photo 19), built in 977 A.D., in the early part of the Sung Dynasty. It is octahedral in shape and reaches a height of 40 m. The underlying

Photo 18. The splendid bridge at Anchi (China, 600 A.D.).

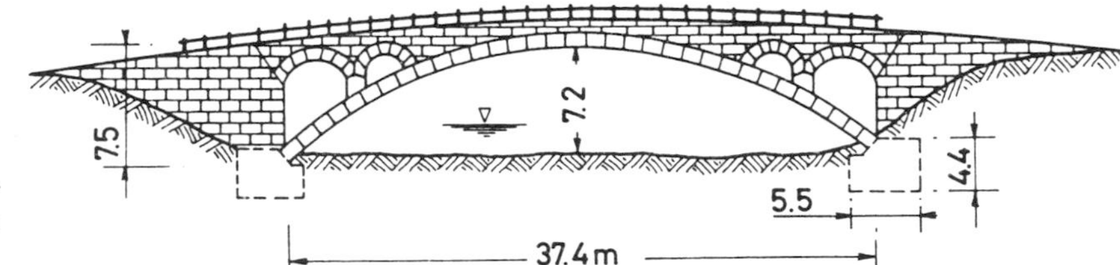

Figure 37. The Anchi Bridge: a longitudinal section after Lu Zhao-Jun (personal communication, 1984).

Photo 19. The Pagoda of Longhua, built in 977 A.D.

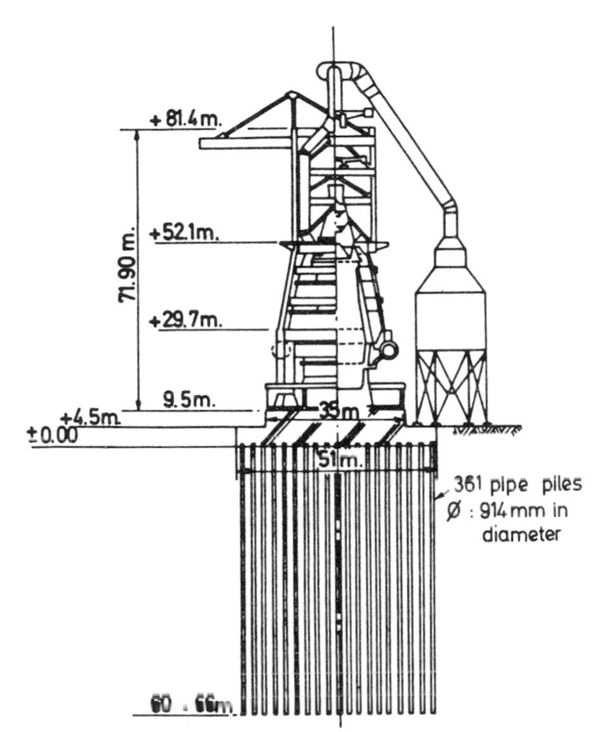

Figure 38. Modern version of the technique used at Longhua: the 35000 ton Japanese blast furnace of Ohgishima, with its 361 pipe piles 66 metres long.

soil is of poor quality, with thick layers of soft clay extending down to a depth of 30 m. According to the China Civil Engineering Society, the allowable bearing pressure on the soil for constructions in the vicinity does not exceed the low figure of 0.8 atmospheres and numerous constructions exceeding this limit have become severely distorted or have collapsed. How is it, therefore, that this forty-metre pagoda is still standing and undamaged, without the least sign of rotation? According to the Museum of Shanghai, the foundations are of brick laid on a wooden raft which in turn rests upon wooden piles at very close spacing. The peripheral piles have a cross-section of 14×18 cm and are set only 8 to 10 cm apart; in present-day terms, this would be called a piled raft, the origin of which therefore dates back over a thousand years.

It is true that, before the Chinese, we find cases of small piles being employed under a monolithic foundation, but the Pagoda of Longhua is apparently one of the first examples of the use of both piles and raft on a large scale.

The same technique was adopted in 1973 by the Nippon Kokan K.K. for the construction of the heaviest blast furnace yet (35000 tons) (Figure 38) on a poor site reclaimed from the sea on the island of Ohgishima (Ishihara 1977). In this case, the 361 tube piles, 66 m long and 91.4 cm in diameter, were made of steel rather than wood. The structure has already successfully withstood several earthquakes, including one with a magnitude of 7.4.

* 6 *

A few hidden errors in medieval architecture

Europe did not match the refined skill of the Chinese for a very long time and the early medieval builders who wished to emphasize the importance of their churches by means of towers had their setbacks. In 985 A.D. the tower of Ramsey Abbey, completed only ten years before, cracked and had to be rebuilt on firmer foundations; the same fate befell the West Tower at Gloucester in the eleventh century, the West Tower at Worcester in 1175 and the central tower of York Minster in the thirteenth century.

The eleventh century saw the emergence in Italy of Republics, Grand Duchies and Archbishoprics whose lands stretched over the rich alluvial plains. These new political authorities competed against one another in erecting towers and campaniles to symbolize their power. Many of these monuments collapsed, sometimes owing to structural defects but more often because the highly concentrated loads were badly founded on the alluvial soil. The visible architecture was undoubtedly creative – but little care was taken with the non-visible part and the poor workmanship was a far cry from the skilled techniques of the Chinese several centuries earlier. Later, even Bramante himself, the great architect of the Renaissance, met with a few unpleasant surprises after spending, for some of his works, the money allowed for the foundations on the decoration of the facades.

It is thus hardly surprising that the list of leaning towers, campaniles and minarets scattered throughout the world contains a majority of Italian towers; some of these – the Asinelli and the Garisenda at Bologna, the Ghirlandina at Modena and the famous Tower of Pisa – are still moving (Kerisel 1985). The rate of rotation of the Asinelli is about the same as the Tower of Pisa but the latter leans at an angle four times greater. Hence it is the Tower of Pisa which always attracts the gaze, not only because of the impressive and increasing tilt which has come down to us through the centuries, but also owing to the architectural beauty of its setting (Photo 20).

This tower, about 20 m in diameter and 60 m high, was built in three stages separated by two periods of repose of about a century each. The first stage was begun in 1173, founded on its site without any special precautions. When the first third had been built, the tower tilted 770 seconds of arc towards the north-west; the town waited for a hundred years or so in the hope of being able to continue, but in the meantime the tilt had increased to 2530″. After some adjustments to provide a level surface at the top of the first part, the second third of the tower was built, and it then tilted successively towards the north, the east and the south, with the angle increasing to 6160″. It was not until 1373 that the decision was taken to finish it. Since then it has continued to tilt towards the south and the angle of tilt is now 19230 seconds of arc, that is 5°20′30″ or 9.3%. In inclining successively towards the various points of the compass it has, so to speak, screwed itself into the ground (counterclockwise when seen from the air) and the average settlement is now about 3 m.

The rotation, or tilt, over the last two decades is shown in Figure 39, which was based by the author on the latest figures provided by the Istituto Geografico Militare in Florence (no measurements have been taken since 1983).

More recently still it has been discovered that the Tower has now stopped settling in relation to the site; it is in fact the converse that is happening – for the whole site is now pivoting from NNW towards SSE and is thus tending to facilitate the rotation of the Tower (Croce 1985).

The builder's error is flagrant – if it can be called an error to have built a monument which attracts

Photo 20. View of the Tower of Pisa and Dome.

more visitors than anywhere else. The Tower is too heavy for the softness of the ground; the quality and precision of its base and superstructure have created a monolithic structure which has had a punching effect on the soil, like the impression of a seal in soft wax, and this impression, now three metres deep, has had repercussions on the clayey deposits lower down, which have been squeezed sideways as various soundings have revealed. (Figure 40)

The present problem is to explain why, in these circumstances, the tower is tilting a little more each year with an almost clock-like regularity. The pivoting of the site as a whole in fact accounts for only part of the tilt. No valid scientific explanation has been offered for the rest of the movement and no one is capable of predicting the future course of events. True, 7″ per year on top of the 19230 seconds of arc so far accomplished is hardly a lot – but in the long term?

So long as the Tower of Pisa continued to screw itself into the ground it was possible to understand why it did not fall. Now that it is no longer settling, it would seem that the ground into which the tower is wedged must be growing firmer and preventing any acceleration of the rotation.

But do we actually have the right to speak of 'ro-

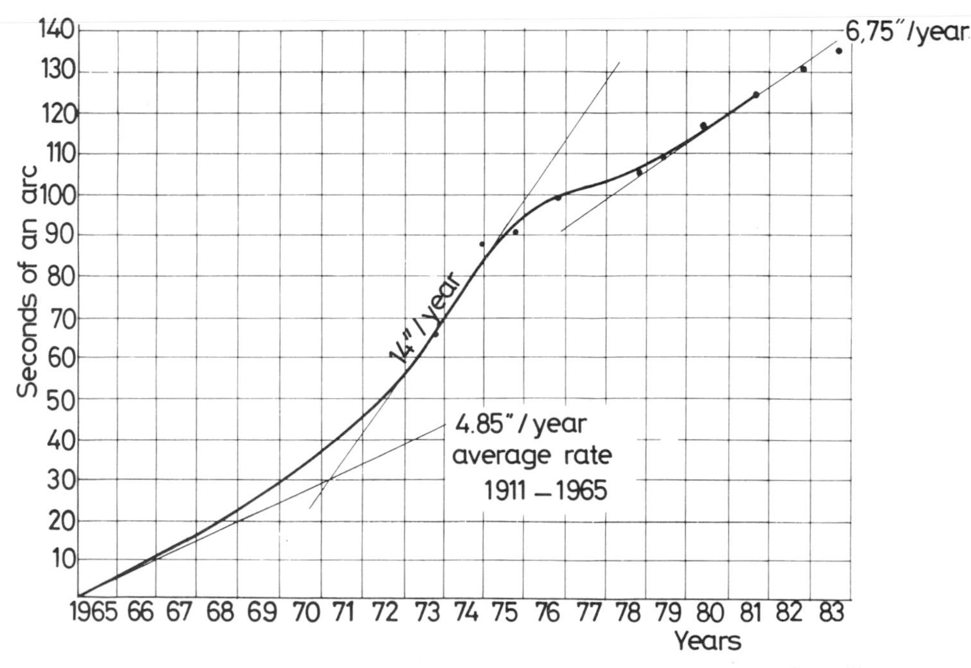

Figure 39. Tilt of the Tower of Pisa over the last twenty years, based on data of the Istituto Geografico Militare concerning the bench marks at its base. The acceleration in 1972-1974 was due to uncontrolled pumping from the water table, now prohibited.

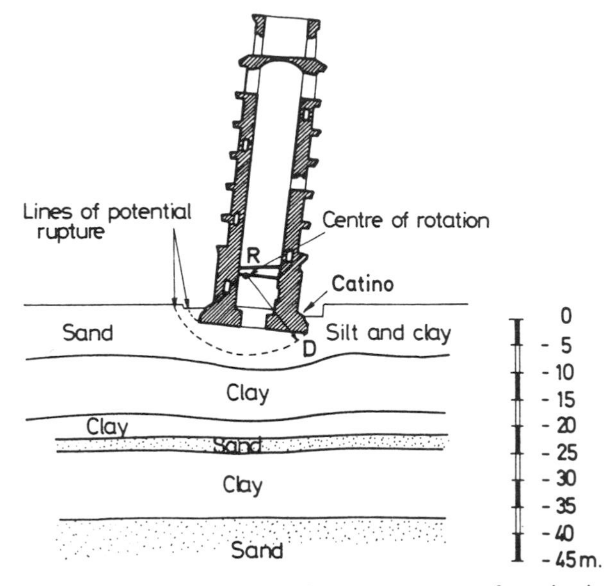

Figure 40. Probable position of the present centre of rotation in the North-South plane.

Photo 21. Engraving of 1782 showing how deeply the columns were embedded in the soil.

tation'? This could suggest that the tower turns about a fixed point or about an axis through this point. When we examine what has been happening since the tower began to tilt in a North-South plane, we find that the movement is at present centered on a point situated at one seventh of the Tower's height and somewhat towards the North (Figure 40). This means that the equilibrium of the Tower depends on the resistance of the soil along the curve of potential rupture (the dotted line in the Figure) and in particular on the shearing strength of the soil in the vicinity of the point D towards which the Tower leans.

An engraving of 1782 (Photo 21) shows the Tower embedded in the soil deep enough to cover the plinths of the columns surrounding its first storey. In 1838, with the intention of bringing these plinths back into full view, the architect Gherardesca had a circular area dug out all round the foot of the Tower. As may be imagined from what has been said, the idea was hardly a success. Water appeared in the deepest part of the hollow, above the point D, and the surrounds of this ring-shaped depression – which the Italians call the 'catino' – had to be cemented (Figure 40).

Meanwhile, the number of visitors who climb it is increasing but the delicately carved stone and columns of the upper part are cracking on the side to which the tower leans.

For centuries now one Italian committee after another has looked into the problem; in 1971 an international competition was organized to find a means of stabilizing (but not righting) the tower. Though many proposals were put forward, no decision was taken; since then, another committee has been appointed with not a single member in common with the previous one, and we are awaiting its conclusions. Nevertheless, in May 1986, the author saw that yet more soundings were being taken at the foot of the tower in preparation for a new project.

Indeed, corrective measures would not be without danger: the tower did not at all appreciate Mussolini's decision in 1934 to carry out exploratory soundings close to its periphery, and signalled its displeasure with a marked increase in its rate of tilt: in the three decades 1918-1928, 1928-1938 and 1938-1948 this rate was 43, 80 and 44 seconds of arc respectively. The stabilization of the Tower is as delicate an operation as the treatment of an elderly invalid by a doctor who is forbidden to sound his chest but who knows for sure that his patient reacts alarmingly to every medicine.

These errors affecting some outstanding monuments rest the exception in Italy and must be seen alongside astonishing successes like the building of Venice on a lagoon – a city of carved stone constructed upon a dreadful soil. This exalting challenge required an adaptation to the soft soil conditions, the acceptance of settlements up to a certain limit together with devices to offset them, judicious pauses in the construction process, careful calculation of the risks involved and ingenious techniques that were both inexpensive and uncomplicated.

The Venetians were always ready to take advantage of the past and build on any ground preconsolidated by the weight of older buildings. Wherever earlier constructors had left their mark no special foundations were laid to support the internal walls of a new more spacious dwelling, whose outer walls alone were provided with new foundations carried on wooden piles (Figure 41). Because of its stiffness and great resistance to traction this wood was also widely used horizontally inside the outer walls to distribute the loads from the stonework.

In order to overcome the sometimes large differential settlements between the internal and outer walls, the cross-beams rested freely on their vertical supports with the ingenious system of adjustment to the walls shown in Figure 42.

An open-framework style of architecture employing arcades and peristyles, that is to say with the loads carried by isolated columns, would seem today rather audacious in such difficult soil condi-

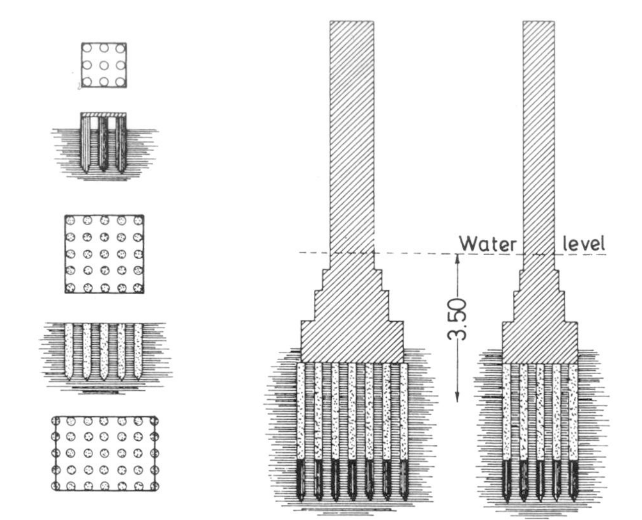

Figure 41. Foundations of the outer walls for a building at Venice. (after Creazza.)

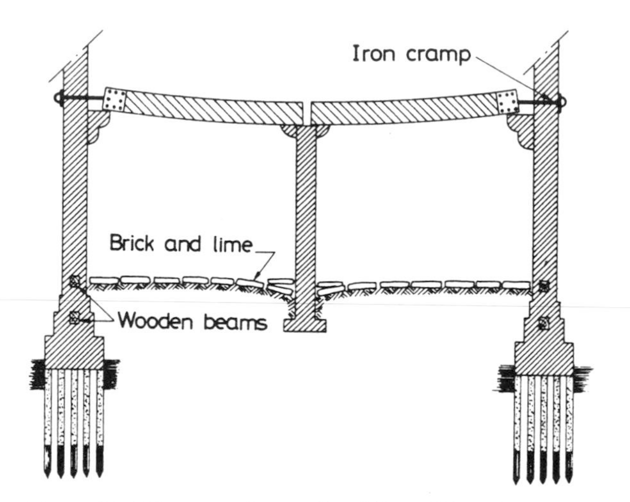

Figure 42. Adjustment devices for differential settlement.

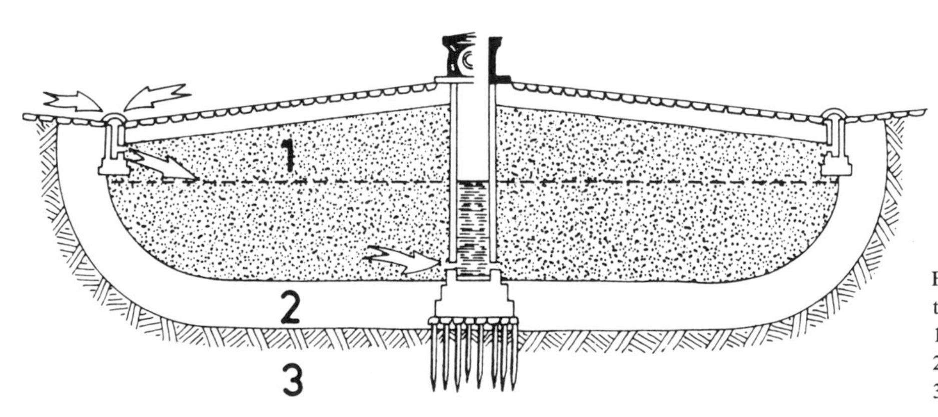

Figure 43. Venetian underground tank
1. Filtering Sand
2. Clay
3. Natural soil

Photo 22. Venetian well.

tions. However the Venetians not only appreciated beauty but also understood how best to arrange the sequence of construction to achieve their ends. Thus for the architraves of the columns they used heavy wooden beams which would deform during construction as the building settled; once the greater part of the settlement had taken place the rest of the deformation was taken out by an additional course of masonry. The same imagination is apparent in the choice of materials for paving the floors, which were sufficiently flexible to follow the settlement of the ground (brick and lime mixed together with oil).

Since the water in the lagoon was polluted, the Venetians collected the rainfall on the roofs of their houses and led it to underground storage tanks (Figure 43). The base of the tanks was curved and was covered by a layer of clay 1 m thick. Placed within the soft Venetian soils a shape like this would normally have been unstable; the solution was to fill it with sand. This ensured stability and served to filter the water, which percolated towards a central well, whose decorated top could be seen above ground level (Photo 22).

Thus the art of preconsolidation, the adjustment and correction of deformations and the patience to wait awhile when necessary made it possible to build this city of art. In a later chapter we shall see the dangers threatening Venice today.

* 7 *

The rampart, the vault and the dome: The itinerary of an idea through several millennia

THE RAMPART

In the early part of his sedentary existence, man was constantly brought up against the same problem: the building of solid structures with sheer faces – defences against invaders or platforms to support temples or palaces – and thus the necessity of containing the lateral thrust. We have already seen how the Sumerians solved this problem by incorporating mats and cables which countered these forces. An alternative approach soon emerged, the wall reinforced with counterforts, an idea which was elegantly carried into effect by the Attalids at Pergamum (second century B.C.) for the retaining wall of the terrace on which the temple of Demeter stood. This wall is 85 m long and 14 m high, with internal and external counterforts evenly spaced along its length (Figure 44). Later, in his 'De Re Architectura', Vitruvius would give a clear statement of the principles involved:

'A series of supplementary walls should be built... to form the shape of the teeth of a saw or of a comb: by this means the earth is broken up into compartments and cannot push on the wall with such a great force' (Figure 45).

Throughout the Roman Empire, and particularly in Spain, the builders of dams followed the advice of Vitruvius. The ruins of two ancient dams of

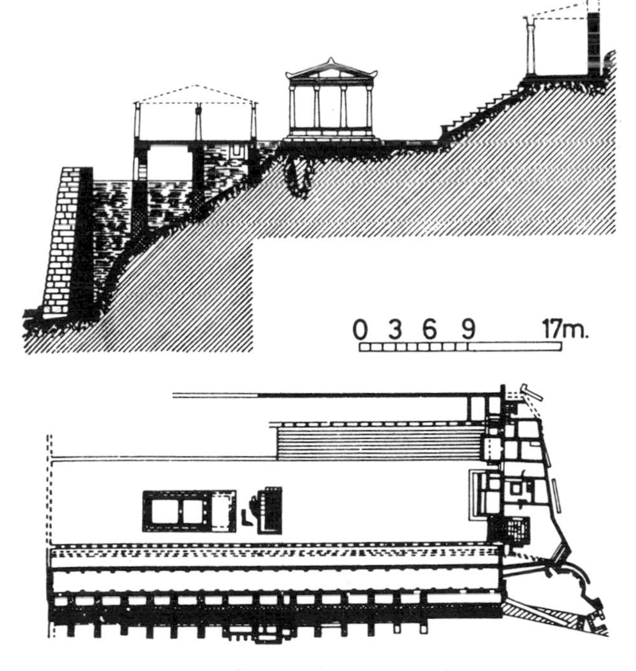

Figure 44. Vertical and horizontal sections of the retaining wall for the terrace of the Temple of Demeter at Pergamum, built in about the second century B.C. by the Hellenistic dynasty of the Attalids.

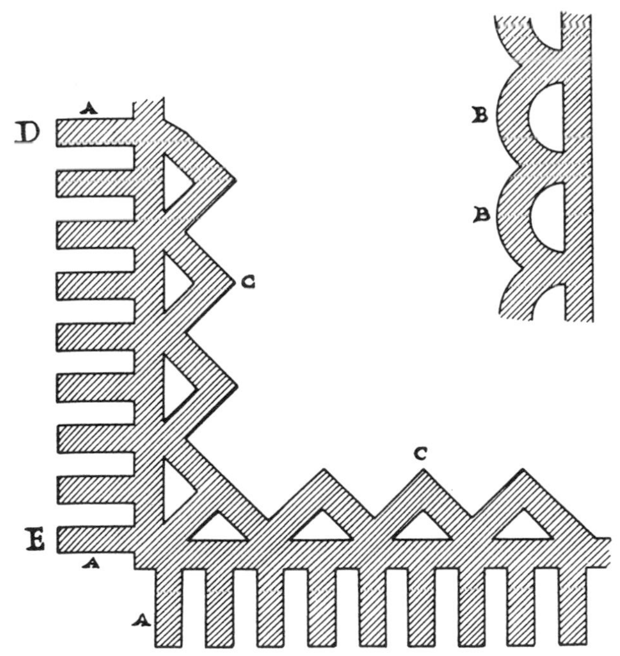

Figure 45. The teaching of Vitruvius on the structure of walls (from De Re Architectura, Book VI, Chapter 11). Illustration by Perrault (1674) of the combs proposed by Vitruvius.

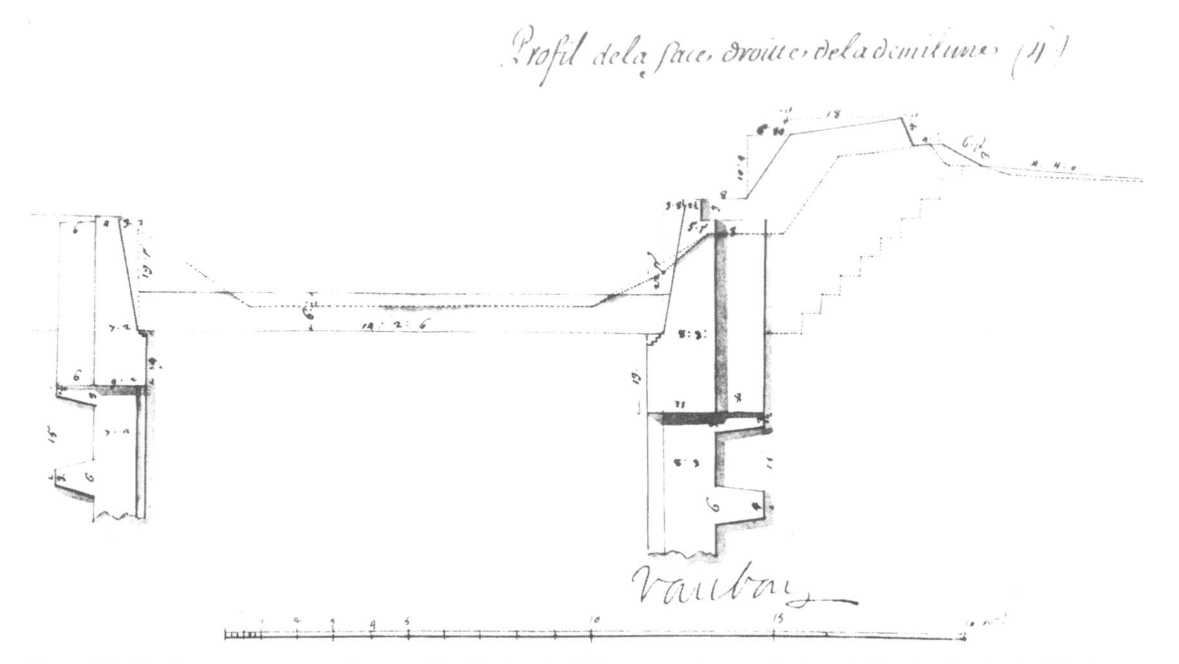

Figure 46. The Pergamum approach as used by Vauban in 1699: two retaining walls founded in clay for the fortifications at Ypres. Photograph of original plan in Archives du Génie at Vincennes.

Photo 23. Modern version of the Pergamum approach: multiple arch dam in reinforced concrete. One of the first dams employing this technique, built on the River Sélune by Caquot around 1920.

earth and masonry at Esparralejo and Consuegra (Fernandez 1961) bear witness to the as yet timid application of the principle on which modern multiple arch dams are constructed.

The French military architect Vauban could not improve upon the approach used by the builder of Pergamum for the defences of the forts he constructed on the orders of Louis XIV at the end of the eighteenth century. After he had built nearly 150 he published 'Tables' with a view to standardizing the spacing, length and thickness of the counterforts, but having analysed the failure of the fortifications at Ypres, erected before his time on a substratum of fat clay, he was forced to admit that

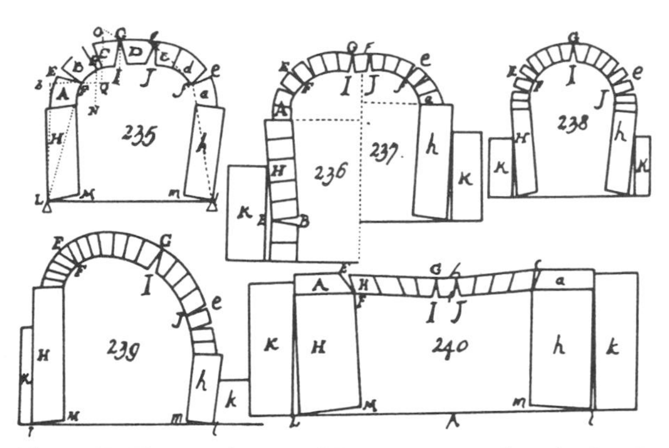

Figure 47. The experiments of Danyzy presented to the Royal Society of Montpellier in 1732.

his rules were only applicable to average soil conditions. The walls had slid forwards and tilted backwards, breaking the joint with their counterforts at the base. This was the first sign of the weakness in the system and was due to the masonry's lack of tensile strength. Vauban solved the problem by considerably increasing the depth of the wall's foundations in the ground at the bottom of the ditch so that the toe of the wall would meet the passive resistance of that soil (Figure 46).

The introduction of reinforced concrete at the beginning of the present century would facilitate such applications of the buttress technique without the danger of separation between wall and counterfort and produce great purity of line in the modern multiple arch dams (Photo 23).

THE VAULT

The sites on which many churches stand conceal the rubble or remains of earlier churches. From the ruins of a chapel rose up a higher and more spacious house of worship, with the older building providing a partial foundation for the new one. Thus the Romanesque or Norman churches gradually replaced those of earlier Christianity, whose wooden roofs not only frequently caught fire but had a limited span which restricted the width of the nave. The ambition of the Romanesque period was to construct increasingly wide and high vaults of stone under which the ecclesiastical singing would resonate. But the builders of the time quickly realized that a thick semicircular vault (barrel vault) with a wide span tends to push the walls on which it rests outwards. Though they did not manage to quantify the actual amount of lateral thrust, it would never cease to be a major architectural headache.

As we have seen, Leonardo da Vinci realized that 'a vault comprises two weaknesses which both work for its collapse but which can be transformed into a strength'. Much later, this concise statement of a fundamental truth was well demonstrated by the experiments concerning arches made of plaster voussoirs submitted to the Royal Society of Montpellier by Danyzy on 27 February 1732. Figure 47 reproduces a page illustrating his experiments Nos. 235-240. If the abutment spreads or tilts, the arch flattens: it opens out at the intrados of the keystone (I, J) and at the extrados of the haunches (E, e) and the weaknesses forecast by Leonardo da Vinci are aggravated. A buttress K simply juxtaposed against the wall H does not improve matters. On the contrary it is easy to show that if K and H are thoroughly locked together and correctly proportioned and founded, the vault can bear heavy loads.

The buttress, therefore, was destined to play a vital role in ecclesiastical architecture: if its foundations and structure were correctly designed, the arch or vault could then soar upwards as a token of religious faith.

During the Romanesque period, the buttresses of the churches were often poorly founded, with badly shaped stones thrown anyhow into the pit with an abundance of mortar. It was thought unnecessary to spend a lot of money on such work, which remained out of sight. According to Jules Quicherat 'Nothing is more common than to find that churches built in the eleventh century fell soon after their construction'.

As for the structure of the buttresses, it was quickly noticed that their central part served no purpose and could therefore be hollowed out. The first instance of this development is to be found at the Constantine Basilica in Rome. Later on, things would be taken further by profiting from the construction of aisles to reinforce the vault laterally, thus replacing the buttresses by an arched portico.

Let us see how this approach was applied at Cluny in the eleventh and twelfth centuries. As the influence of this abbey spread its abbots wanted to build wider and higher and to let in more light. Being of a very pragmatic turn of mind, they did not start to erect their famous abbey church, Cluny III – called by texts of the time the 'maior

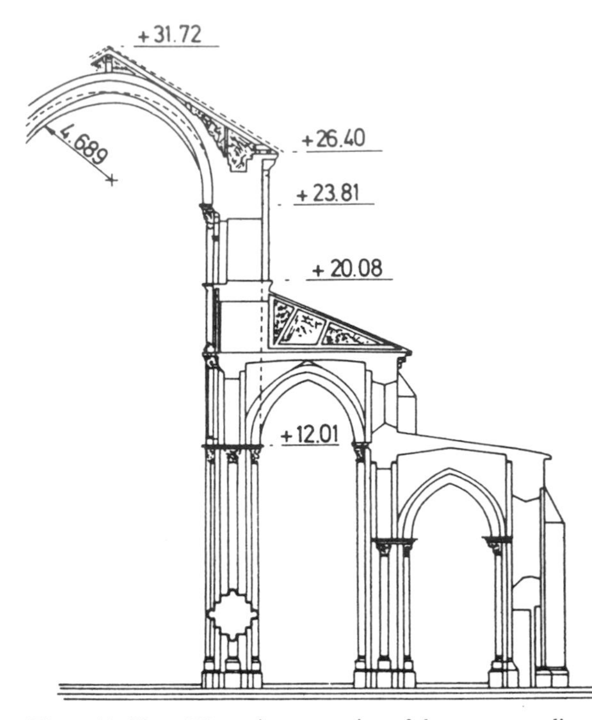

Figure 48. Cluny III: semi-cross-section of the nave according to Conant.

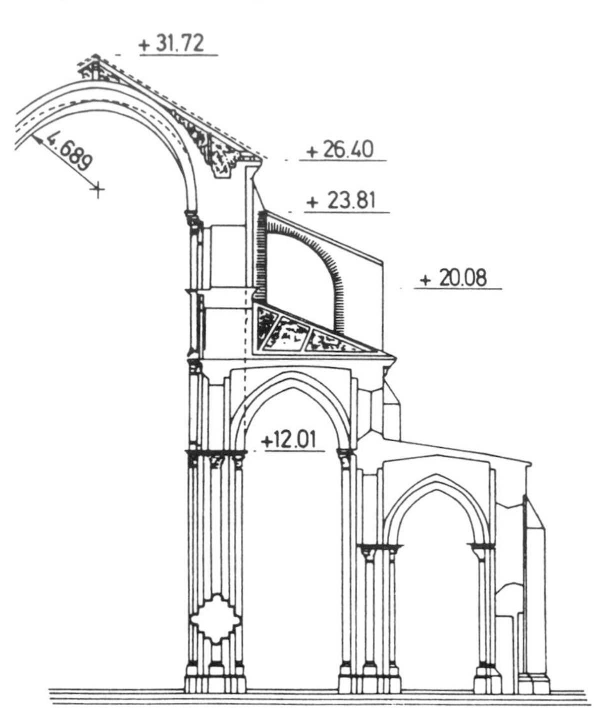

Figure 49. Cluny III: semi-cross-section of the nave after the repairs (1130).

ecclesia' – until they had observed the behaviour of the previous ones, Cluny A, Cluny I and Cluny II. The first two were without aisles but Cluny II had one on each side, the building measuring 13.50 m from wall to wall and having a nave 7.50 m wide. Cluny III would have two aisles on each side and the distance between the external walls would be extended to 39 m and the width of the nave to 12.15 m. The jump in size was audacious, making it the highest abbey church of the Romanesque period, with the crown intrados 29.5 m above the slab floor. Figure 48 shows a cross-section to the right of one of the pillars of the nave: the clerestory was not buttressed until 7 m below the haunches of the vault and, worse still, those seven metres contained many windows (covering about 48% of the clerestory) thus aggravating the vulnerability to lateral stresses. In 1968 the great American archaeologist K.J. Conant reconstituted the abbey, showing the light-filled beauty of the architecture. It would seem to have successfully anticipated the natural lighting of the future Gothic churches while maintaining the Romanesque style.

Alas, the vault collapsed in 1125. To rebuild it the inner aisles would be topped, bay by bay, with a hollowed out buttress (Figure 49) that rose to counter the thrust of the vault near its haunches. The repairs turned out to be exactly what was needed and the abbey church remained in perfect condition until orders were given for its demolition during the French Revolution.

The transverse view of Figure 49 hints at the idea of flying buttresses on two levels, and suggests why it has been said that the Gothic style was to a certain extent born out of the attempts to repair Romanesque churches.

The flying buttress appeared in the second half of the twelfth century, at the same time as the ribbed vault, one of the two other fundamental features of Gothic architecture, which both diminished the outward thrust and simplified its canalization.

But even in the thirteenth century certain builders of Gothic vaults remained faithful to the Romanesque buttress. The problem of reducing the thrust of the vault was interpreted in different ways by the builders; here it is interesting to compare King's College Chapel in Cambridge with the Sainte Chapelle in Paris (Figure 50). The later was built around 1240, about 200 years before the former, but they have more or less the same internal dimensions. The buttresses of King's College

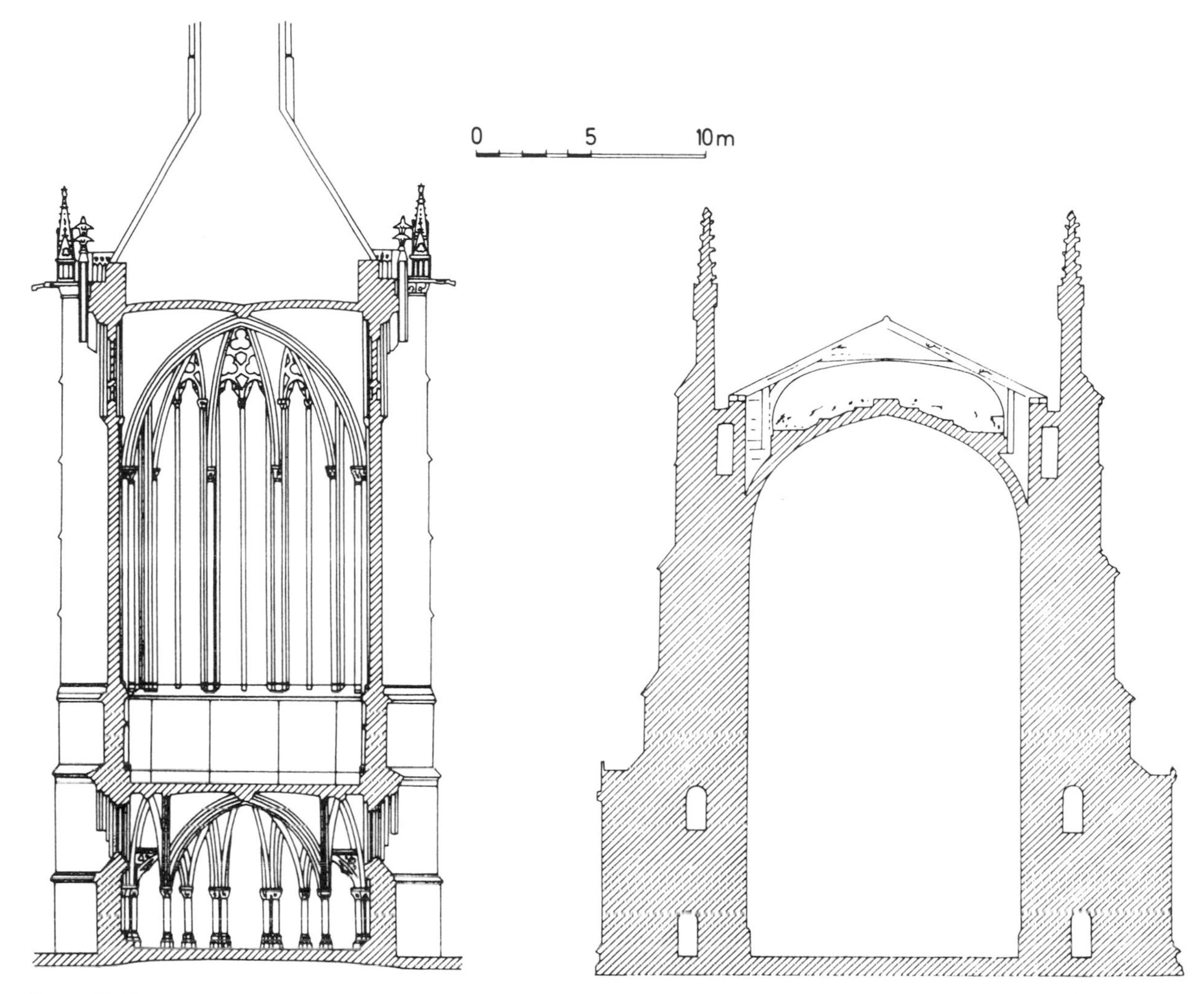

Figure 50. For comparison, cross-sections of two churches from the Gothic period with almost identical internal dimensions: The Sainte Chapelle (left) and the Chapel of King's College, Cambridge. The two diagrams are on the same scale.

Chapel are not only 2.5 m wider but are set much closer to one another. Yet the soil conditions were probably no better for the Sainte Chapelle since the Ile de la Cité where it stands has been submerged several times by floods of the Seine. The Sainte Chapelle's grip on the soil is thus, in comparison, extremely discreet. 'Difficile ex imo' – it is difficult to build on bad soil – and yet that is what Pierre de Montreuil did.

When they came to raise their Gothic naves on a sheer rock, the builders had no choice but to employ the buttress wall and leave us to admire, for example, the boldness and elegance of the buttress architecture rising from the steep slopes of Mont Saint-Michel to provide support for the Merveille (Photo 24).

But flying buttresses in carved stonework soon replaced the solid form and, here too, the result could be dangerous. One example is the choir of Beauvais Cathedral, which marked the apogee of Gothic architecture with its height of 48 m and which had one of its pillars collapse in 1284, twelve years after it had been finished; the transept had not yet been built and the nave never would be. Excavations have revealed that, as in the case of Cluny III, the failure could not have been due to an error in the foundations (Kerisel 1975). Figure 51 represents the section through the choir. The architect had used a double flying buttress and taken every care to make sure that the intermediate pillar O and the outer pillar, which receive the thrust of the vault transmitted by the flying buttresses, were

Photo 24. An audacious version of the Pergamum technique: the buttresses of the Merveille, on the northern side of Mont Saint Michel.

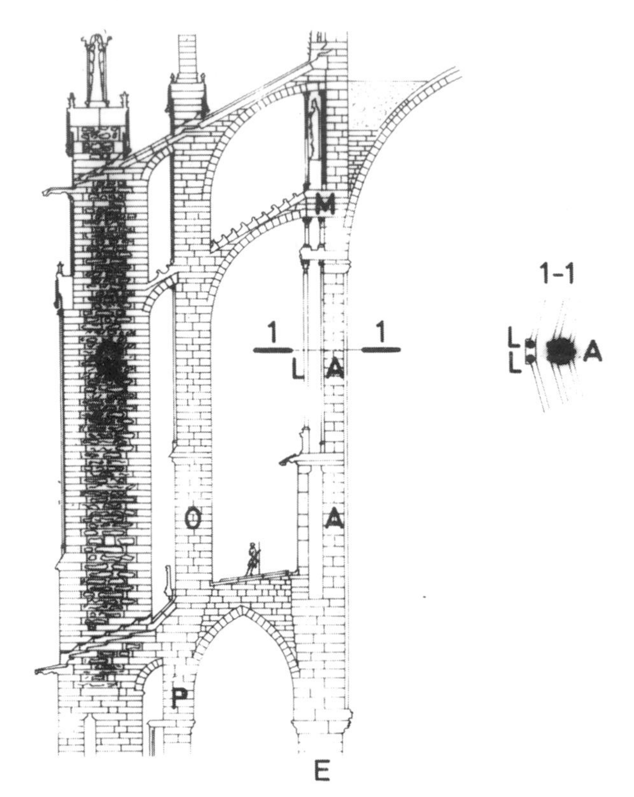

Figure 51. Section of the choir of Beauvais Cathedral, from Viollet-Le-Duc (1868).

heavy enough to produce a well-centered resultant. But he overlooked the fact that a large proportion of the weight of the vault is transmitted to the pillar E. These pillars, which adjoin the nave, were thick at the base but, further up, divided into the pillar A and the twin shackled columns L, as shown in section 1.1. The latter were of well-dressed stone without a central filling and thus much stiffer than the pillar A which had a rubble core. Little by little all the forces were transferred to these slender columns, with the result that they eventually broke. Viollet-Le-Duc (1868) and Heyman (1967) also explain the accident in this way. Subsequently, an additional pier was constructed and the flying buttresses strengthened.

DOMES

In 1743, the dome of St. Peter's in Rome cracked, with the cupola splitting rather like an orange. Like vaults, domes generate lateral thrusts which tend to fracture the drums which support them and to force the tops of the weight-bearing pillars outwards.

Unless these pillars are short and thick they cannot really perform the function of buttresses and the drum itself has to be tied with a chain bond. This is what was done for St. Peter's – and for the dome at Florence from the outset – and for a number of other domes above the middle of the transept of large cathedrals. On the other hand, the dome of the Roman Pantheon profited from the fact that the monument was specially designed for it: it has stood the test of time magnificently because its drum rests on a series of walls which radiate outwards to form bays and act as solid buttresses, thus ensuring the equilibrium of the cupola.

However, it was not so much the question of buttresses that lay behind the mishaps with the big domes but rather the transmission to the ground of the great vertical loads represented by the cupolas and their supporting drums. The dome of St. Paul's, for example, which is the largest in the

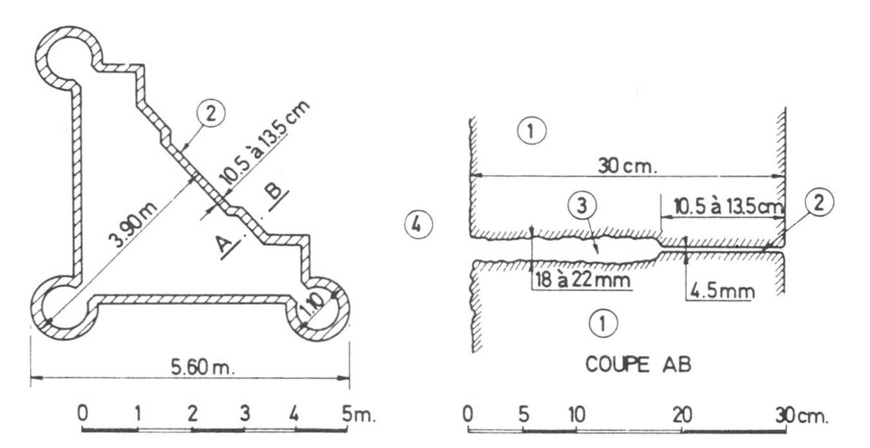

Figure 52. Pillar supporting the dome of the Pantheon in Paris. Horizontal section and detail of joints.

world, places a total load of 70 000 tons on the London clay beneath it but this enormous weight, as well as the outward thrust of the dome, is uniformly absorbed by the eight huge piers or bastions with the result that the whole has progressively but safely settled about 16 cm. On the contrary in some other cases we find the same mistakes in handling the transmission to the ground of these enormous loads as were made with the choir at Beauvais. Here are two examples.

Louis XV entrusted the construction of the Pantheon in Paris to Jacques Soufflot, whose design included a dome supported by four triangular piers with imbedded columns (Figure 52). This dome, although smaller than those of St. Peter's in Rome and St. Paul's in London, nevertheless places a load of 3 100 tons on each of the piers. These piers were formed of a rubble core in plenty of mortar with an outer casing of carefully dressed stone, a long-established practice for their construction. But in the case of the Pantheon, Soufflot committed a number of errors. In the first place, his cut stone was not of the best quality. Secondly, and probably as a measure of economy, the dressed part of the joints, as is shown in the diagram on the right, only extended 12 cm in from the outer face of the stone. The core was highly compressible on account of the abundance of very slow-setting mortar. In other words, the 3 100 tons to be supported by the piers were borne by the narrow outer band that had been dressed, that is to say, by only one sixth of the pier' surface. Cracks appeared as early as 1776 and increased in number as the construction of the dome reached completion. In 1796, the dome was in danger of collapse and had to be shored up; later, after the death of Soufflot, Rondelet, with great perspicacity and skill, diverted part of the load towards the core of the piers by sawing through the facing stone in order to provoke its settlement and thus oblige the core to carry more of the weight.

The pillars supporting the dome of Milan Cathedral had a marble casing on the outside and a core of granite blocks, marble chippings and bricks in a bed of mortar. The cathedral was terminated in 1813 and the stresses gradually concentrated on the marble. Cracks appeared and the situation had become so serious by 1969 that one of the pillars had to be jacketed with reinforced concrete to prevent it from collapsing. Here we find the same weaknesses as with the French Pantheon, though much less pronounced – which explains why the cracks took a long time to appear.

Thus we see how the idea of buttresses, so elegantly expressed by the architect of Pergamum, has wound its way through two thousand years of military, monumental and ecclesiastical architecture, taking a variety of forms and occasionally, especially in ecclesiastical architecture, paying the price of excessive boldness. But the necessary correctives were found so that the strength of which Leonardo da Vinci spoke could overcome the inherent weaknesses of vaults and enable them to stand up to the passing centuries.

Part Two
The last two hundred years

* 8 *

Coulomb ushers in the scientific approach

As we have seen, throughout the long period of Romanesque and Gothic churches their builders had to rely on 'warnings': when something collapsed or went wrong it meant that they had overstepped certain permissible limits. This is what happened to the vaults of Cluny III and Beauvais, to the domes of St. Peter's in Rome and St. Sophia in Istanbul and to other illustrious buildings. They knew that certain forces could bring down vaults or crack domes and that the earth could topple retaining walls or slide down into cuttings, but little headway had been made towards the scientific understanding of the forces involved.

It is not surprising then that one of the major problems facing scientists in the eighteenth century was that of determining the minimum force needed to withstand the thrust of arches and vaults. At the same time the deep ditches which Vauban incorporated in his defensive systems made it vital to understand the nature of the thrust of the earth against the increasingly high retaining walls.

These two problems appear to be quite different in kind and it was the achievement of Coulomb, in a Memoir presented before the Royal Academy of Science in Paris in 1773, to show that they both called for the same mathematical method for the calculation of maximum and minimum forces, an approach that opened the way to our present-day concept of upper and lower limits. The Memoir has been translated into English under the title 'On an application of the rules of maximum and minimum to some statistical problems relevant to architecture' by J. Heyman (1972), who also provides an excellent critical study of the history of these two fundamental problems from the eighteenth century to the present day. In 1973, the eighth International Conference on Soil Mechanics and Foundation Engineering, held in Moscow, marked the importance of Coulomb's paper by celebrating the bicentenary of its first publication (Kerisel 1973).

Coulomb, of course, is much better known for his work on electricity and magnetism than for this study, which was the main focus of his earlier life, and, as Gillmor's biography (1968) relates, cost him much physical suffering.

In 1757, at the age of twenty-one, Charles Augustin Coulomb was elected a deputy member of the Société Royale des Sciences of Montpellier and it was there that he probably met Danyzy who, twenty-five years before, had presented the experiments concerning arches to which we have already referred (Figure 47). It is therefore very likely that Coulomb was already aware that arches collapse owing to the formation of hinges. At 26 years old, he graduated from the Mézières School of Military Engineers and two years later, as a young officer, he was sent to Martinique, a Caribbean island that the King wished to fortify. His plans for a fort dominating the entrance to the harbour – known as Fort Bourbon – were accepted and he was put in charge of building it. He had taken with him to Martinique a copy of Bélidor's treatise for military and civil engineers but became dissatisfied with the rules given for calculating the dimensions of the walls. He did not agree with those published by Vauban either and decided that the first thing to do was to reconsider the problem experimentally. As he wrote later, 'I have often come across situations in which all the theories based on hypotheses or on small-scale experiments in a physics laboratory have proved inadequate in practice: I did all the necessary research . . .'. He would spend eight years in the heat of the Tropics, where fever and dysentry rapidly led to the repatriation of his superiors and eventually undermined his own health.

THE THRUST OF EARTH

Before Coulomb, it had been recognized that the

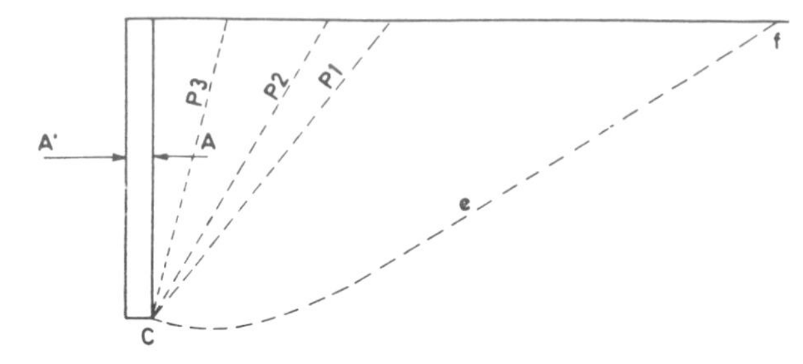

Figure 53. A corresponds to a rupture plane (P1, P2, P3, etc.) passing through C. Clearly, the opposite force A' has to be far greater than A if it is to rupture the soil behind the wall (Cef).

collapse of walls was due to the development of a surface of rupture in the earth but its exact location and form were either unknown or were fixed a priori in order to construct a theory.

The first merit of Coulomb was to discover by experiment that this surface was a plane and then, by calculating the forces exerted by all the possible planes (P1, P2, P3 in Figure 53), to identify the one which exerted the maximum thrust A on the back of the wall.

Having found the plane of weakness and the corresponding A, his second merit was to distinguish clearly between the thrust A (corresponding to earth in an active state) and the thrust A' which has to be applied to the wall from the outside inwards in order to provoke a line of rupture C e f (Figure 53) (corresponding to earth in a passive state). Though he did not determine the shape of this line (which is not straight), he showed that A' is superior to A.

His third merit was that his calculations allowed for both friction and cohesion along the line of rupture. Up until then little attention had been paid to this cohesion (i.e. the inherent strength of the soil).

In 1777, Coulomb published a memoir on friction which, although dealing with substances other than earth, described a number of soil properties which have been confirmed by recent research: that friction at rest increases with time but, on the other hand, decreases towards a residual value when the relative movement continues.

The value of scientific studies often lies not so much in what they actually advance as in the way they stimulate others. This is what happened with Coulomb's Memoir, as Heyman demonstrates: it inspired further studies throughout the nineteenth century (Rankine, etc.) up to the work of Boussinesq (1882), who provided complete equations (but not integrable) for the problem of active and passive pressure, giving values for any slope of the soil and any batter or overhang of the wall, including the friction on the wall's internal face. These equations have been utilized to produce tables of figures for active thrust and passive resistance (Caquot and Kerisel 1948). More recently, Sokolovskii has developed a powerful mathematical tool, which reproduces or refines the results deduced from the equations of Boussinesq. The first question raised by Coulomb, with respect to the thrust of earth, has thus found a final answer in the work of Boussinesq and Sokolovskii.

THE THRUST OF ARCHES

'I have used the rules of maximum and minimum' wrote Coulomb, 'with the object of determining which would be the true sections of rupture in arches which were too weak, and the limits of the forces that could be applied to those of given dimensions'.

Figure 54, taken from the Memoir, shows how Coulomb sought all those sections mM in which the resultant of the thrust at f (assumed to be horizontal) against the section Ga mM and of the weight of that section is directed either through M, thus tending to draw that part of the arch (Ga mM) downwards or on the contrary through m. This

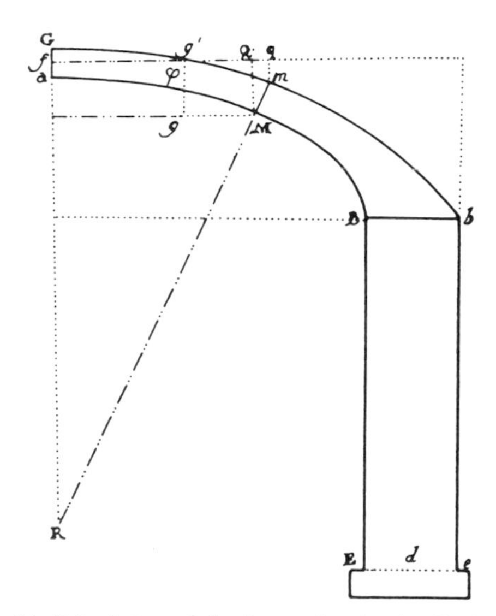

Figure 54. Calculation of the thrust of arches by Coulomb (Figure 14 of Memoir).

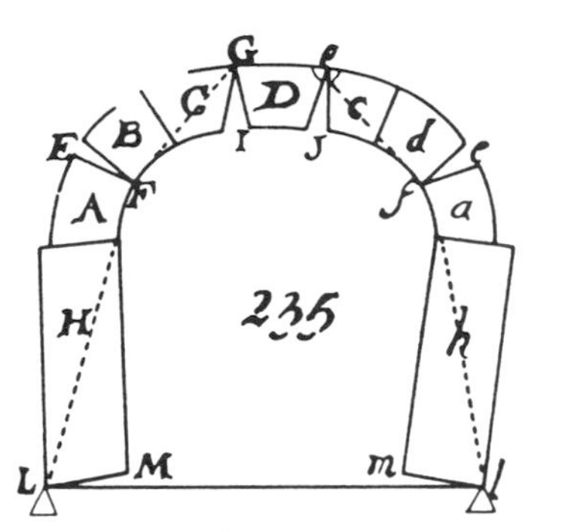

Figure 55. The probable line of thrust (dotted line) in Danyzy's experiment N° 235. (1732)

enabled him to discover that the thrust was at a minimum when the point f was near the top of the section and at a maximum when it was near the bottom (Figure 56).

In Danyzy's experiment 235 (Figure 55), for example, the minimum thrust H is found at the point G shown in Figures 54 and 55 and in diagrammatic form in Figure 56 (a). This has always been the crucial problem for the builders of churches since this thrust can force the support outwards. All that has to be done is to make sure that the counterforts or flying buttresses are able to contain it. The case in which H is at a maximum (Figure 56 (b)), pushing the supports inwards and the keystone upwards, rarely occurs in practice.

Though Poleni (1748) anticipated Coulomb in showing that the line of thrust should pass entirely within the masonry of the arch, it was Coulomb who quantified the problem and proved that the sole mode of failure was the formation of hinges at the intrados or extrados. It is quite true that the problem is purely a question of geometry and that the actual strength of the stone can be disregarded when the pressure exerted by the thrusts lies within the section of the arch, because the masonry can absorb it easily. But if hinges form this is no longer correct since the thrust is then concentrated on a very small surface.

NEW PROBLEMS ARISING WITH THE INDUSTRIAL AGE

The development of industry quickly confronted Coulomb's successors with fresh problems.

In the nineteenth century in particular, the construction of railways and navigable waterways called for deep cuttings which, when the soils disturbed happened to be clayey, were frequently the cause of sudden landslides. In 1846, Alexandre Collin, having collected and analyzed data on some of these accidents, published a treatise on 'Landslides in Clay'. This treatise appears to have been quickly forgotten since a careful reading of it would have dissuaded Ferdinand de Lesseps, thirty years later, from rashly undertaking to build the Panama Canal.

Collin described the sliding surface as ressembling a cycloid (Figure 57) and pointed out that along this surface the clay was polished. He showed that the slip surface developed gradually as the cohesion of the clay weakened, and he devised experiments to measure this cohesion in terms of its consistency. This led him to suggest certain ways of avoiding such slides: vertical 'counterforts' in the form of gravel or stone-filled trenches which would perform the dual function of drain and buttress (a new version of the Pergamum counterforts combined with the drainage mats used for the ziggurats). But the contribution of Collin was not recognized by the scientific community until 1956, when an English translation of his study showed twentieth century civil engineers (Petterson 1955, Fellenius 1927) that their theory of rotational slides was in fact very similar to his approach.

Collin had perceived the fundamental role of the cohesion of the clay and that this cohesion could be destroyed by water. Water problems of another kind, linked to the need for pumping from the sandy aquifers to permit the growth of cities, prompted further scientific progress. Ten years after Collin's treatise appeared Darcy's well-known study on the filtering of water in sand for the supply of the city of Dijon. This study still forms the

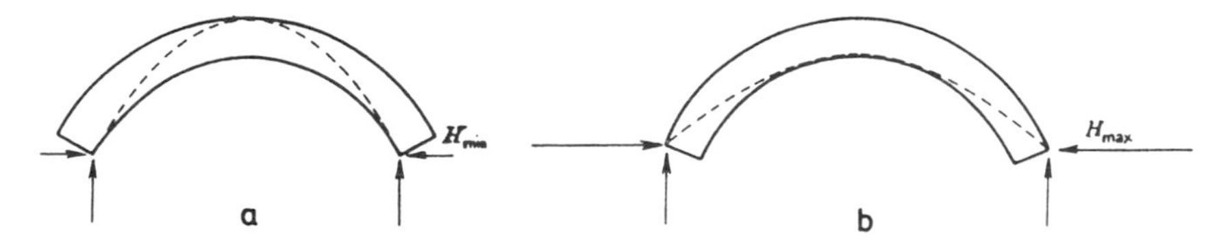

Figure 56. Positions of the line of thrust in the case of H min. and H. max.

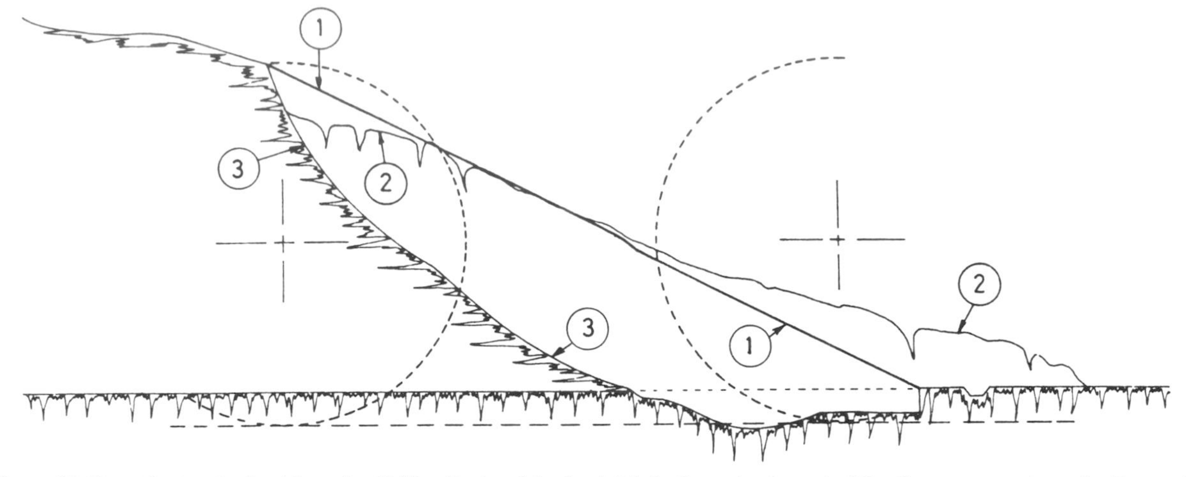

Figure 57. From the treatise by Alexandre Collin. Study of the landslide in the embankment of the Cercey reservoir on the Burgundy Canal in 1846. The slipline (3) is a cycloid which is the movement of a point on the circumference of a circle (dotted lines) which rolls on a horizontal plane. (1) represents the original soil profile and (2) the profile after the landslide.

basis of hydrogeology though it was actually published only as the final annex to a long memoir for the municipality of Dijon which was mainly concerned with the administrative and financial aspects of a supply system for drinking water. Darcy, who wanted to prove that the public fountains would be adequately supplied, described his experiments on the filtering of water through a layer of sand with a thickness L and announced that if the difference in head after the filtering was h (Figure 58), the filter rate was directly proportional to h and inversely proportional to L. This was the starting point for the science of hydrology.

With Darwin (1883) and Reynolds (1887) there was another big step forward in understanding the properties of sand. Darwin showed that the coefficient of friction increases with the density of the packing of the grains and Reynolds discovered that if densely packed and totally saturated sand is sheared, the volume increases and the water pressure drops.

But it was only between 1910 and 1930 that important progress was made in the understanding of clays. The landslides in the Culebra Cut during the building of the Panama Canal and the failure of the upstream slope of the earth dam at Charmes (France) when it was drawn down in 1909 led to serious study of the problem. In 1911 Atterberg defined the water content associated with changes in state from solid to plastic to liquid, and in 1914 Frontard, with reference to the Charmes slide, showed that the pressure of the water in the clay of

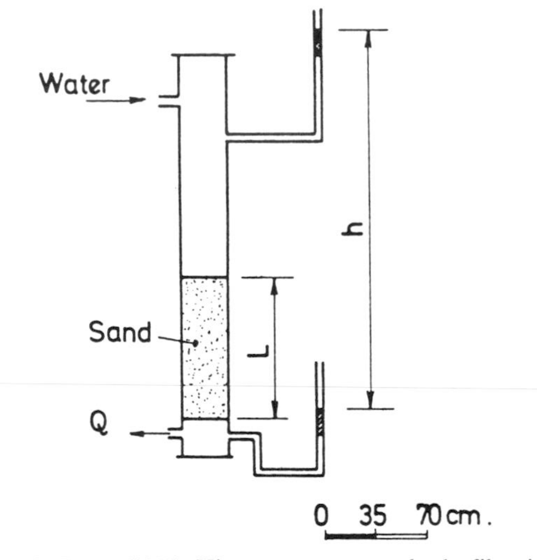

Figure 58. Darcy (1856). His apparatus to study the filtration of water through sand.

the dam depended on that in the reservoir.

However, the fundamental contribution was made by Terzaghi (1925), who demonstrated that the water in saturated clays, which is incompressible compared to the clay structure, increases in pressure when the clay is loaded (Figure 59). Then, as the water drains out, the pressure slowly drops and is gradually transferred to the clay structure. Before this transfer, the clay behaves as if it possessed cohesion, with friction coming into play only with the onset of the transfer. Terzaghi showed how this process (now called consolidation) evolves over time and depends on the compressibility and

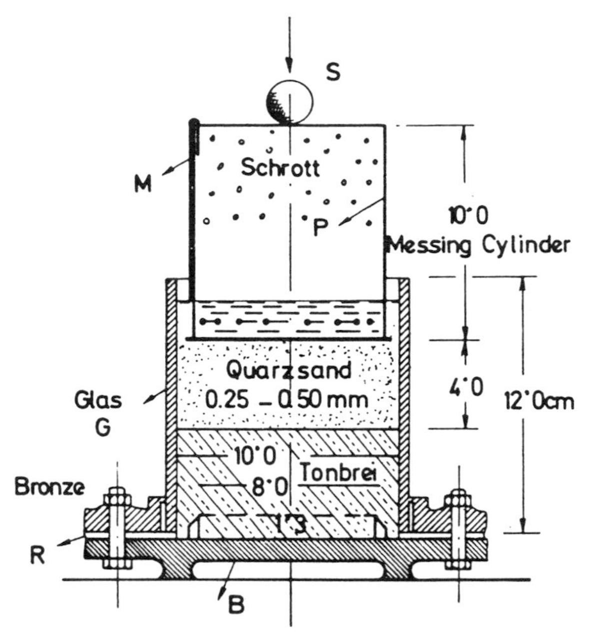

Figure 59. The oedometer: a device invented by Terzaghi for studying the consolidation of saturated clays under a constant load. The water in the clay (tonbrei) rises into the sand (quarzsand)

(very low) permeability of the clay by measuring these two factors under laboratory conditions (Figure 59) and determining the mechanical strength at the beginning and end of the transfer.

By 1927, then, the foundations for a science of soil mechanics based on sound physical principles had been laid. The measurement of soil compressibility now made it possible to apply the theories of Boussinesq (1885) in order to calculate the settlement of soils under the weight of constructions. What was the weight limit for constructions before the soils collapsed? It was at this point that the problem recalled the work of Coulomb on the second equilibrium, the 'maximum limit' (see A' in Figure 53): if one imagines the wall in Figure 53 as a horizontal foundation and the external force A' as the weight of a very tall building, the soil beneath the foundation will collapse when the load reaches a certain limit.

Accidents in Japan caused by certain earthquakes, as well as oil prospecting, now focused attention on the transmission of vibrations through the soil and revealed in the substratum the existence of loosely-structured sands in a metastable state which could liquify when subjected to vibrations. Such sands could be particularly dangerous if they lay beneath high-security structures such as nuclear power stations.

Lastly and most important of all, experience would make it clear that the most refined mathematical studies in soil mechanics are meaningful only if they take account of precise and detailed geological data, especially in the case of heterogeneous soils. This was the painful lesson of the Panama Canal at the beginning of the century, which led to much greater research on the mechanical and physical characteristics of soils (cohesion, friction, etc.).

Such then, briefly summarized for the non-specialist, were the principal landmarks in the scientific understanding of soil mechanics over the last two centuries. More detailed information may be found in Skempton (1985) and Peck (1985).

* 9 *

The Panama Canal: An ocean-to-ocean cutting with multiple side effects

Since the time soil mechanics became a science in its own right, a lot of precious information has come from the accounts of near disasters in engineering. Who now recalls the construction of the Panama Canal? The first attempt led to a financial crash in France, whose engineers set out to dig the most difficult of the world's canals, a deep and unstable cutting through a hostile landscape. It proved a bottomless pit for the small subscriber before eventually turning out, for the Americans who subsequently took over, an extremely profitable investment. It is one of the greatest marks that man has left on the earth's surface, a major monument of international civil engineering – and hence of particular interest to us.

The first blow with a pick-axe was given about a hundred years ago, on January 1 1880, yet in Europe the centenary of this important moment has slipped by without a single publication despite all the lessons to be learned. The Americans, on the other hand, who in 1904 took up the torch abandoned by the French in 1889, have recently brought out several works (McCullough 1977, and Anguiloza 1980) recalling the long history of this gigantic undertaking, which was eventually completed in 1914.

When the project for a canal through the central American isthmus was born, France was admired for two outstanding successes: the Two-Seas Canal and the Suez Canal.

The Two-Seas Canal, completed under Louis XIV in 1675, was 240 km long and linked the Atlantic to the Mediterranean via the Neurouze gate, a pass 192 m above sea level. Along the stretch between Toulouse and Neurouze, Pierre Paul de Riquet (1604-1680), who designed and built it, had had to contend with some difficult stretches which forced him to ease the earthen slopes on each side but, always on the spot, he succeeded in finishing the job. Though, later on, Vauban had to make a few improvements here and there, foreign observers generally agreed that the bold undertaking had been a remarkable success.

Suez, on the other hand, was mainly the outcome of delicate negotiations, a triumph for the extraordinary diplomatic skill of Ferdinand de Lesseps. The technical side did not involve enormous difficulties: the soil, almost entirely sand, was on the whole easy to dig and the topography, with the highest point only 15 metres above sea level, was highly favourable; the climate was hot but dry and the few problems encountered by the contractors were quickly bowled aside by the dynamic leadership of de Lesseps.

The success of the Suez venture blinded people to the real qualities of de Lesseps: he was called 'The Great Engineer', the one thing he was certainly not. According to Bunau-Varilla, he in fact despised engineers and regarded them as impractical people. The truth is that he eventually stopped listening to them, a fact which accounts for his behaviour at the inter-oceanic congress he convened in 1879, just before the job of digging through the Central American isthmus was started. He manipulated the 136 delegates in Paris so as to get them to approve a project for which no serious studies had been undertaken since he had already made up his mind to adopt the shortest route and a lockless canal at the mean level of the two oceans. This route involved a cut 109 m deep through the backbone of the isthmus, the Culebra Cordillera (Figure 60). But that was not the only contrast with Suez. Even more important was the local environment and geology of the isthmus, which turned out to be almost unbelievably varied and difficult. When the Americans took over, they rightly criticized the lack of studies by the French, but they too had their problems. It was indeed a far cry from Suez: the

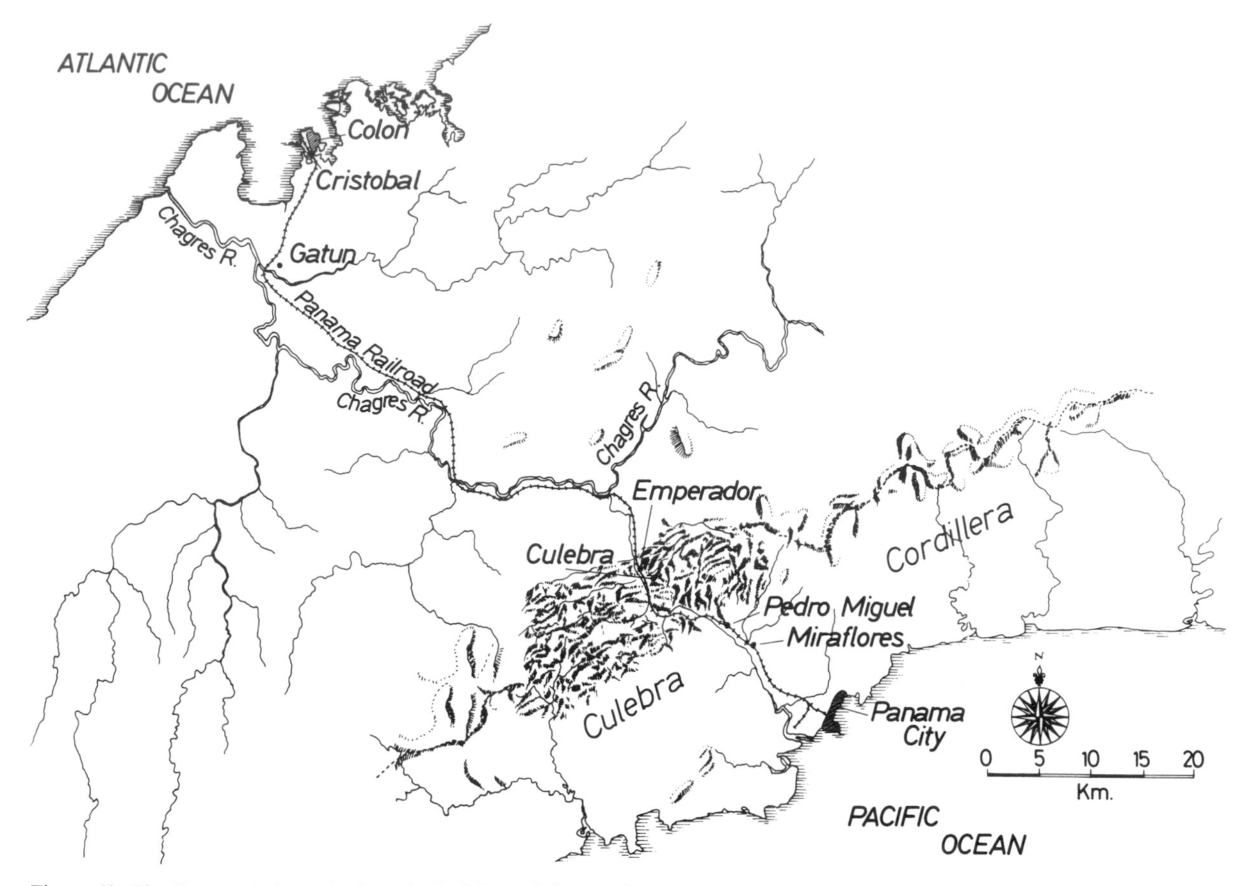

Figure 60. The Panama isthmus before the building of the canal.

Panama Isthmus had only recently emerged from the sea and was composed of soft clayey formations that were unstable and liable to slides. Nor could it be said that the Congress gave no warnings. Godin de Lépinay, who had built the railway from Cordoba to Vera Cruz and explored the site chosen for the canal, knew about the poor quality of the subsoil and the torrential force of the local rivers; with Gustave Eiffel, he was one of the few delegates to vote against the project, proposing an alternative scheme for a canal on several levels, with a lake on the Atlantic side, formed by a dam at Gatun coupled with locks, to absorb the floodwater of the Rio Chagres and supply the locks (Figure 61). This plan had the immense advantage of reducing the depth of the Culebra Cut. But de Lesseps remained deaf to such pleas and died ten years before the Americans, in 1904, returned to the project of Godin de Lépinay.

To give himself a clear conscience and create a climate of confidence among the subscribers, de Lesseps tried – with success – to get the Academy of Science to give a favourable opinion regarding the route and the geology of the terrain. In 1880, the learned assembly, which was usually most circumspect in its reports, unwisely produced two favourable opinions: 'the design and workplans of this enterprise are worthy preparations for an undertaking that will benefit the whole of mankind' (p. 274). And then, on the basis of its examination of a few samples taken from the shallow cutting dug for the existing railway, it gave its views for the Culebra Cut (p. 275): 'For some 25 km the rock sides should be sloped back at 1 metre horizontally for every 4.25 m of rise'. In fact, however, as we have said, the channel was not through rock but through a tangle of formations, some of sedimentary origin with a high proportion of clays and others with a volcanic history which took on the consistency of soap when wet. The almost vertical sides envisaged by the Academy would actually end up as relatively gentle slopes with the result that the volumes of

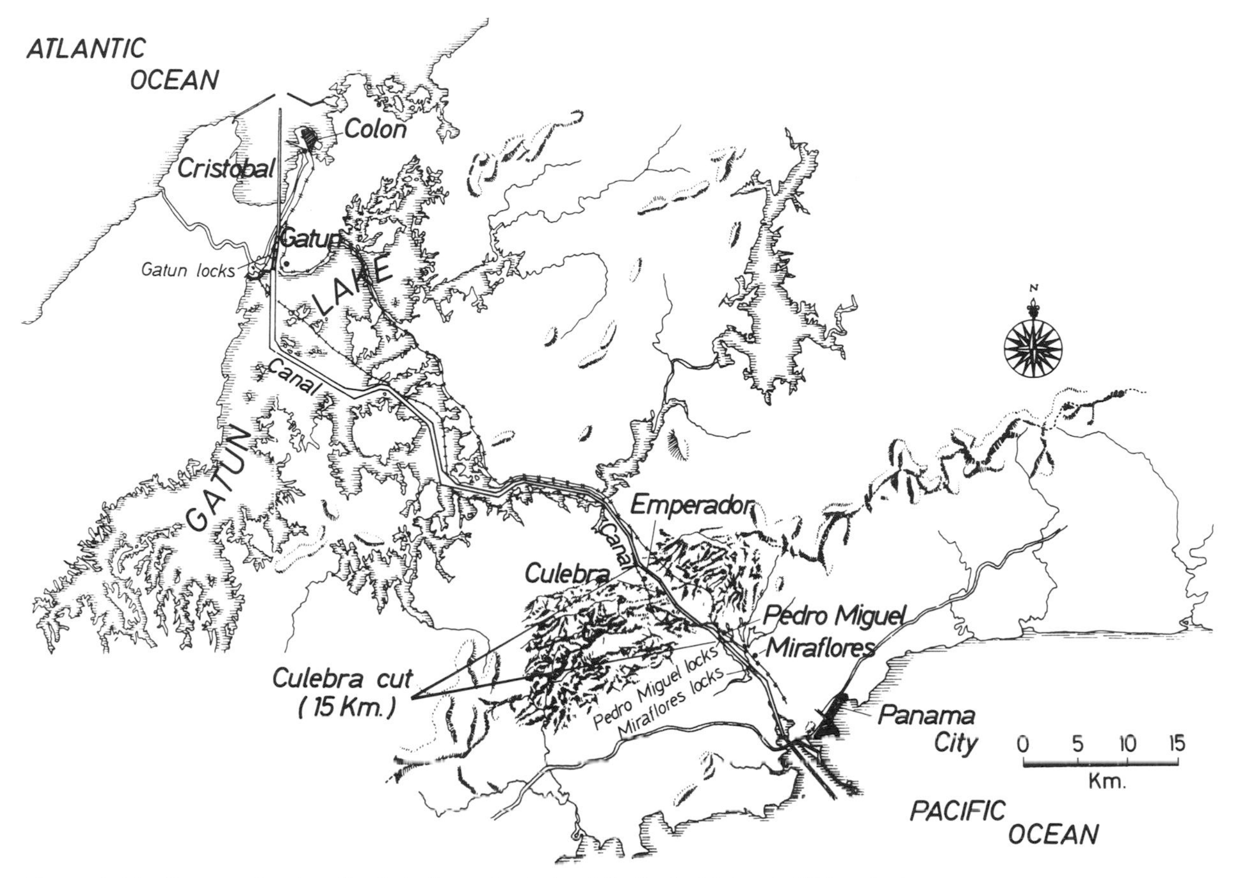

Figure 61. The Panama Canal as finally built on the basis of Godin de Lépinay's 1879 project.

material to be excavated greatly exceeded the orginal estimate: by the time the Canal was opened in 1914, the French and the Americans between them had excavated around 170 million cubic metres of material instead of the 46 million announced by de Lesseps.

Nor had the environment anything in common with Suez. Suez averaged 22 mm of rain a year whereas Colon averaged 3000 and Panama 6000; in addition, the jungle of the isthmus harboured yellow fever, which killed at least 10000 workers, engineers and members of their families.

The true geology and the complexity of its history became apparent as soon as the digging started.

The narrow strip of land joining North to South America had emerged from the sea and sunk again at least four times, the most recent uplift (which is still continuing) having taken place only 10000 years ago. At Culebra, there are a large number of secondary faults and the formations are composed of soft black shales, marls and clays recently deposited when the isthmus was under water, and these are interbedded with layers of lignite sandwiched among sandstones, small basaltic dykes and, on the east bank, a basaltic cap on a meta-tuff known as Gold Hill (the astute prospectus for potential investors stated that this dome-shaped hill had got its name because it rested on a base containing veins of gold!). The 70 km of the canal passes through no less than seventeen different rock formations, six major geologic faults and five major cores of volcanic rock.

Under the weight of the upper layers, the clays had been transformed into argillites, i.e. weak clayey rocks, but the interbedded layers of sand contained water which, in conjunction with the heavy rainfall, initiated a softening process. The result was that the slopes became increasingly unstable as the cut deepened and quickly gave way in landslides until they achieved a gradient of about one in four.

Here is the account of a contemporary eye-

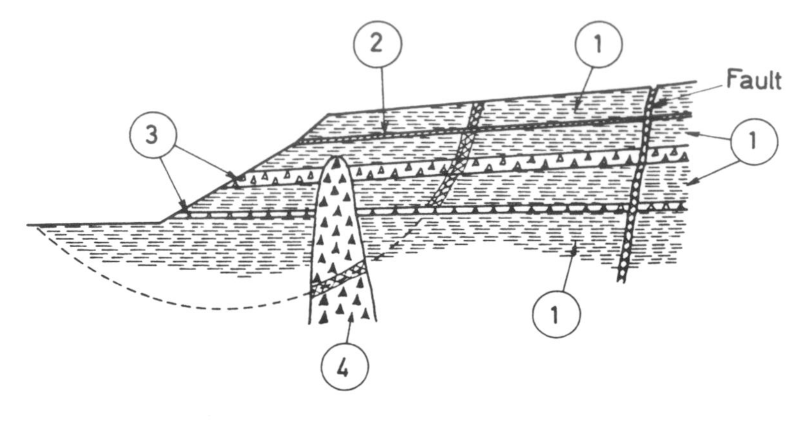

Figure 62. The second type of slide at Panama.
1. Weak clayey rock with beds of mud and tuf.
2. Shales and lignite.
3. Conglomerate and cemented gravel.
4. Andesite lavas.

witness, an American engineer called S. W. Plume, who visited the site in 'the French time': 'The whole top of the Culebra hill is covered with boiling springs. It is composed of a clay that is utterly impossible for a man to throw off his shovel once he gets it on. He had to have a little scraper to shove it off [. . .]. The slide of clay quickly spilled over the site's railway track, and a new one was built on top. The same thing happened a year later so that where the present track is there are two underneath. One morning my house, which had been about 150 m from the cut, was suddenly at its very edge.' (McCullough, p. 166).

The more one dug, the more slides there were and the more material was left to be excavated. The slides were in fact of two types:

1) those occuring in the rainy season were caused by the softening of the not entirely impermeable clayey formations which slid on the underlying layers of rock (which sloped towards the cut), like snow sliding down a roof;

2) others were caused by the fracture at a certain depth (Figure 62) of the rocky formations, owing to the lateral thrust engendered by the excavation: i.e. the elimination of the passive resistance which the excavated soil had previously provided. Through fissures in the rock there developed cycloids, those rotational slides described by Collin some fifty years earlier in connection with the Cercey dam and the deep cuttings dug for the French canals between 1825 and 1850 (Figure 57). These slides gave forewarning at the surface: huge crevasses would appear at the top and a few weeks or months later the whole mass would give way.

Meanwhile yellow fever struck. The Chief Engineer Dingler lost in succession his daughter, his son, his daughter's fiancé and finally his wife. He gave up in 1885, in a year during which, of a hundred new arrivals, at least twenty would die and only twenty of the others would remain strong enough to work normally. Dingler was succeeded by Hutin, who gave up after only a month, and then by Boyer, a young and highly talented engineer, who died of the disease a few months later.

The work was halted in 1889. Up to the last minute Ferdinand de Lesseps had been shamelessly continuing to feed Paris with false news. In 1888 his son Charles, at a banquet for contractors, declared in his name: 'The Canal will be opened in 1890. We have altered our work programme. Since the mountain would not come to us we have gone to the mountain. We have simplified the problem and reduced the volume to be excavated. I drink to the health of the French workers.'

To take just the Culebra Cut, the French withdrew after excavating only 14.5 million cubic metres, little realizing that the total amount dug out of this cut alone would eventually amount to 75.5 million cubic metres – and then not for the sea-level canal wanted by de Lesseps but for the present lock canal which passes through the Culebra Cut at 25 metres above sea level.

One man lived through the entire history of the canal, a French engineer named Philippe Bunau-Varilla. It was he who promoted the secession of Panama from Colombia to become an independent State. As Panama's Minister Plenipotentiary he negotiated the re-opening of the site; he was a controversial figure but generally admired by the Americans for his practical turn of mind and his efficiency (Anguiloza 1980).

In 1904, then, the Americans adopted the Godin de Lépinay project, with a dam at Gatun and a big lake to the east. They brought in numerous geolo-

Photo 25. Culebra Cut. Looking south. December 1910.

Photo 27. Culebra Cut: another landslide on the east bank. View looking south. September 19, 1912.

Photo 26. Culebra Cut: landslide on the east bank. February 10, 1911.

Photo 28. Culebra Cut, September 21, 1918. Looking east. An island has emerged as a consequence of a landslide. Photos 27 to 28 courtesy of the Library of the Panama Canal.

gists but, despite further studies and much more powerful equipment, they had many setbacks, including enormous landslides in 1911 (Photo 26) and 1912 (Photo 27). In 1913 the canal was partially filled with water so that work could proceed more economically with powerful dredgers, but the presence of the water encouraged further slides owing to seepage through the fissured rocks. Soon after the canal was opened to traffic, small islands emerged here and there and in 1915 it was completely blocked by landslides from both banks (Photo 28). In fact traffic was occasionally interrupted by other slides right up to 1933.

The library of the Panama Canal preserves the memory of the difficult and tragic struggle of the French and American pioneers, whose problems can be imagined from Photos 26 to 28 of its archives. From the viewing point at the top of the Culebra Cut one can see the ships pass by 100 metres below, symbol of a busy international economy while, to the informed onlooker, the black reflection of the water on which they ride seems like a pall of lead drawn over the lives of so many of its builders and the despair of so many small investors.

What are the technical lessons to be learned from Panama? Where de Lesseps is concerned the answer is clear: a completely inadequate plan, without prior study of the soils and consideration of the area's geological history; disregard for the scientific work of Alexandre Collin; and an abusive reliance on the Suez experience. As Godin de Lépinay had so rightly told de Lesseps in 1881, 'to take the same approach in such contrasting environments is an outrage to nature and a negation of the engineer's duty.'

But the difficulties encountered by the Americans show how exacting the problem was. Their numerous setbacks would lead to a radical transformation of soil mechanics and geology: in particular, sampling at depth, in situ and laboratory tests came to be used much more extensively.

FINANCIAL AND POLITICAL REPERCUSSIONS

It is rare for a scheme to leave such financial, political and social scars. In his fall, de Lesseps brought down Eiffel, and a number of politicians ended up with severe prison sentences. Eiffel, as we have seen, had voted against the sea-level project of 1879; in 1887, however, when Ferdinand de Lesseps at last realized that a sea-level canal was impossible and adopted one with locks, Eiffel, urgently persuaded by the financier Baron de Reinach, accepted a highly lucrative contract under which he undertook to complete the work on 10 enormous locks in 30 months. After a series of audacious successes he was at the pinnacle of his career, with the Eiffel Tower then steadily rising above Paris in record time. His decision to join the project made a great impression on public opinion and renewed the confidence of investors. The French Parliament authorized the sale of lottery bonds for 720 million francs (1 000 million francs had already been subscribed in 1880). But the bonds did not reach the expected price and the financial barons, 'with Jewish-sounding names' such as Lévy, Crémieux, de Reinach, Herz and Aron, stepped in. When the Panama Company went bankrupt an investigation revealed corrupt practices and public opinion blamed it all on the financiers. An anti-semitic campaign gathered momentum, eventually culminating in the celebrated Dreyfus Affair.

* 10 *

An architecture that defies the heavens

A NINETEENTH CENTURY BEGINNING: THE EIFFEL TOWER

On the eve of the Universal Exhibition in 1889 the Eiffel Tower boldly pointed its 320 metres towards the sky. Though no longer the tallest structure in the world, it is still one of the most visited; but those who take a lift to the top are usually quite unaware of how it is rooted in the ground. When a person looks at it, he generally directs his eyes upward; let us for once reverse this temptation and regard it from the top down.

The architecture of the tower was engendered by the wind. In 1890, its creator made this clear: 'The first principle of aesthetics in architecture is that the essential traits of a monument must be in perfect harmony with its purpose. What were the conditions I had to consider in designing the Tower? Wind resistance. Well, then, it is my claim that the curve made by the four leading edges of the monument, which was determined by calculation, conveys a profound sense of strength and beauty, as the girders rise from an enormous and most uncommon span at the base and gradually narrow until they reach the top'.

But it was important for Eiffel to estimate correctly the full violence of the hurricanes that the Tower would have to withstand at some time or other in its existence. The structure was designed for a wind of 300 kgf per square metre and Eiffel also checked it against the almost equivalent hypothesis of a crosswind increasing from 200 kgf/m^2 at ground level to 400 kgf/m^2. Even with its open steelwork, the force exerted on the Tower is no less than 2554 Tf (2185 Tf in the second hypothesis), or not less than 30% of the dead weight of the Tower (7341 tons). This was a most unusual percentage and the situation was made worse by the fact that the centre of action of this horizontal force was located at 84.90 m from the ground (98.55 m in the second hypothesis), thus exerting a considerable overturning moment.

From the numerous measurements by anemometer at the top of the Tower, it has been deduced (Dettwiller 1969) that there is the statistical probability of a wind of 234 k.p.h. once per hundred years. The manifest stability of the tower has confirmed the accuracy of the relationship, imagined by Eiffel at a time when so little was known about aerodynamics, between windspeed and the pressures exerted on the exposed surfaces.

The distance, 101.40 m, between the pillars (Figure 63) is greater than the height of the centre of wind action and this attenuates the effect at ground level. As a matter of fact, each of the pillars has four separate blocks of foundation, one for each girder. Figure 64 represents a vertical section of block C of pier 1. The steel girder transmits its downward thrust, at an angle of 54° to the horizontal, to an elongated block of masonry aligned with the diagonal of the Tower, the weight of which somewhat straightens the thrust of the girder. The dimensions at the base (5 m wide and 15 m long) are so calculated that the maximum wind pressure will not increase the stress corresponding to the dead load by more than 50%, giving a total pressure of 50 Tf/m^2 at the outer lower edge of the foundation block.

All the equipment added to the Tower since it was built has increased its weight by about 30%, while the winds to which it has been exposed have been somewhat less violent than expected; the design figure for the overturning moment therefore remains valid.

But the second requirement was that the soil be able to withstand the rather high pressure mentioned above. The subsoil of the site is mainly composed of a very thick layer of plastic clay overlying

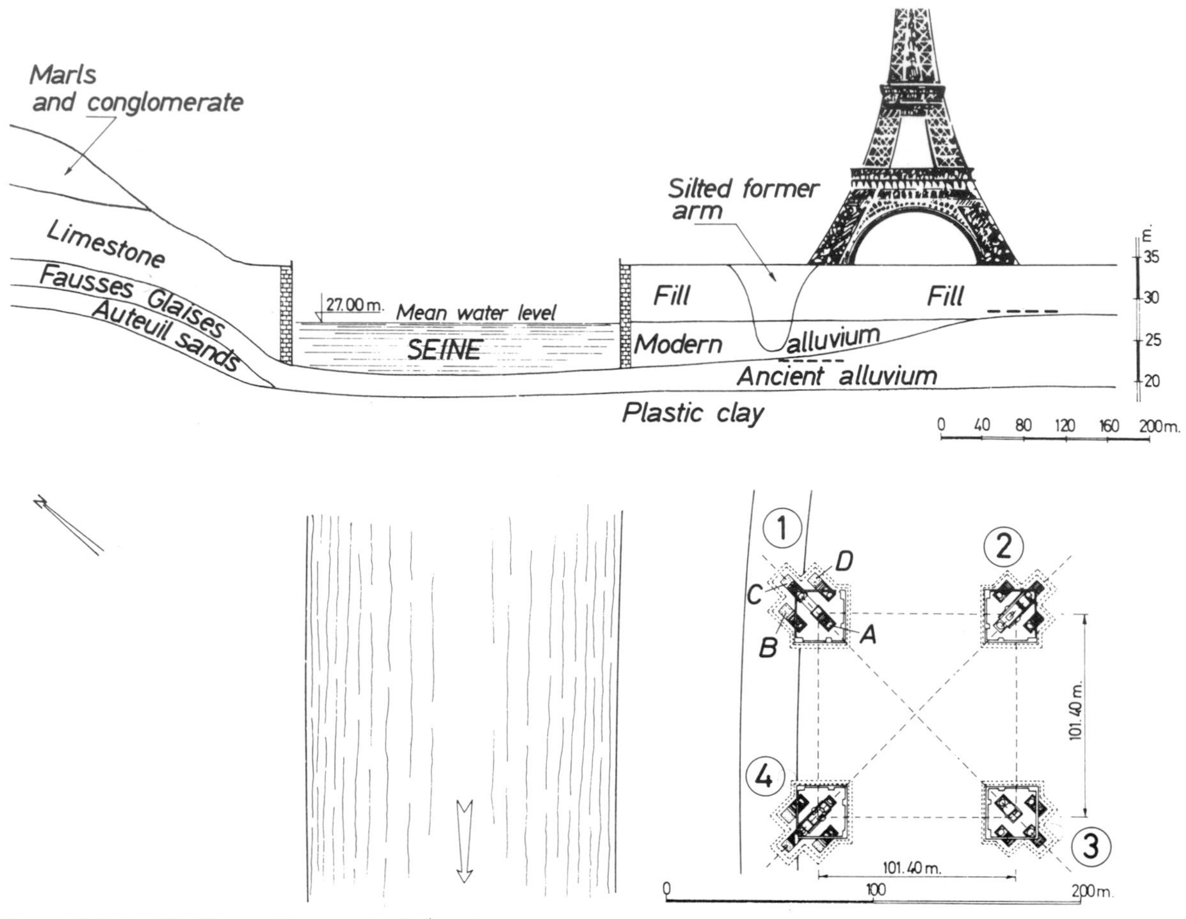

Figure 63. The Eiffel Tower: Above, geological section - - -: foundation levels. Below, diagram of foundations.

chalk; above the clay lie successive deposits of two types of alluvium (Figure 63). The first type, made up of ancient alluviums or very compact sand and gravel, is quite thick (7 m) at a certain distance from the Seine, and in particular under piers 2 and 3, which were founded without difficulty above the water level on this substantial bed. Nearer the Seine, however, these ancient alluviums had been eroded and covered by a second type of more recent alluvium which is of very poor quality. This was the situation under piers 1 and 4; to make matters worse, they happened to be located on the edge of a former arm of the Seine, which had become silted up and been filled in a hundred years before.

Had the plastic clay been laid bare, thus making the whole undertaking impossible? If not, did there remain a thin bed on top of it that could carry the load? How was one to know?

'In order to make an accurate survey of the difficult terrain on which the river-side piers would have to be founded, we had begun with the usual soundings, but the results were so full of uncertainty that we could not be satisfied. What conclusions could one reasonably base on the examination of a few cubic decimetres of excavated soil, more often than not diluted by water, and brought to the surface by the scoop?' In view of all this uncertainty, Eiffel decided to explore the soil by another much more unusual means: working with compressed air, he drove an iron tube 200 mm in diameter down to below the water level and managed to discover that the plastic clay was dry and overlaid by a bed of modern alluvium that was 3.37 m thick. He found, however, that the alluvium was not homogeneous but included blocks of limestone, chlorite and fallen rocks no doubt originating from the limestone buff opposite (Figure 63).

As all the plots for the Universal Exhibition had already been attributed, it was impossible to change the site of the tower. Could Eiffel rest his

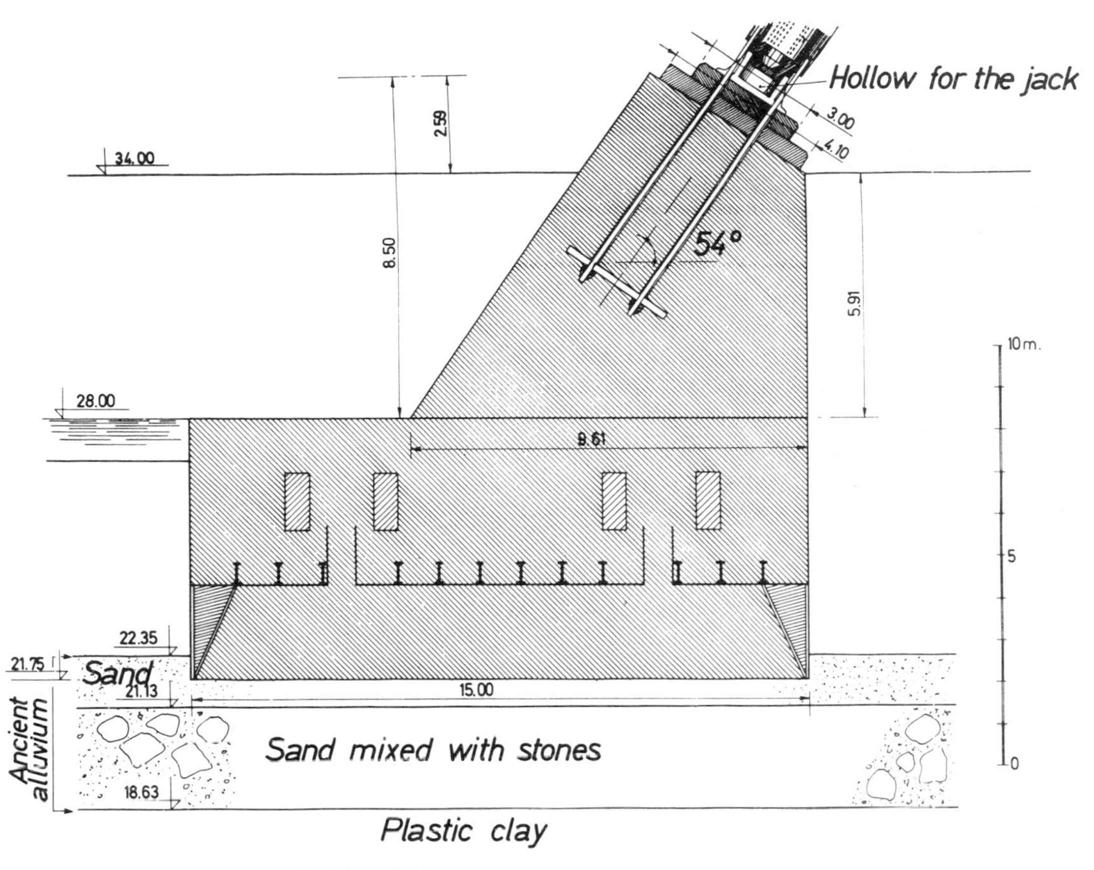

Figure 64. Section of foundation block C for pier 1.

foundation on this small heterogeneous bed? He had just before been the butt of bitter criticism by 'art lovers' who were against the building of the Tower. They had even addressed a lampoon to the Director of the Universal Exhibition: 'We, writers, painters, sculptors, architects, and passionate lovers of the hitherto unspoilt beauty of Paris, protest with all our might and all our indignation... against the erection of that monstrous and pointless Eiffel Tower which public scorn... has already baptised the Tower of Babel'.

Eiffel had to make sure that the Tower lay on foundations that would not collapse yet would allow the iron superstructure to sway during a hurricane. He had calculated that winds of that violence would provoke a horizontal movement at the top of only 1.04 m provided that the soil remained firm. Only 32 years previously, however, the Alma bridge (1855-1874), which was but a few hundred metres up the Seine and whose piers had been founded on a bed of ancient sands some 3.50 m thick overlying plastic clays known as 'fausses glaises', had settled nearly a metre. Was Eiffel going to repeat the Tower of Pisa and be the laughing-stock of those who had written the lampoon?

Concluding that the bed of alluvium and the clay were both of better quality than at the Alma site, Eiffel took his decision – which time has proved to be the right one. One cannot help admiring his boldness and judgment for, at the time he built the Tower, he had none of the laboratory testing equipment available to present-day engineers.

All the eight foundation blocks of piers 1 and 4 on the Seine side of the Tower were completed, from 28 m above sea level down to 21.75 m, by means of compressed air (without a decompression chamber) and the remains of old masonry, tree trunks, mud and all sorts of debris were penetrated without difficulty.

There is a persistent legend that the Tower is still resting on jacks placed beneath it to compensate for the settlement of the soil. The truth is that Eiffel did indeed make use of jacks, but to facilitate the assembly of the steel girders which emerge from the piers; in Figure 64 one can see the hollows in which they were positioned. But once all the adjustments had been completed, the bottoms of the main girders were sealed into the foundation blocks and the jack-holes filled in.

In all probability, Eiffel never dreamed that so many people would visit his Tower. Yet he wanted to avoid any inconvenience for those who did. A small paragraph in the newspaper *Le Temps* for 27 March 1887 reads: 'Messrs Becquerel and Mascari, Members of the Institute, paid a visit to the Champ de Mars. They recognized the quality of the work being done on Mr. Eiffel's foundations: the metal structure of the Tower has been brought into contact with the water table so that visitors will be in no danger from the effects of atmospheric electricity'.

As a matter of fact, Eiffel, to be absolutely sure of this, had buried between the solid blocks of foundation, beneath the level of the Seine, some long cast-iron pipes the upturned ends of which were connected to the ironwork of the Tower by electrical conductors.

ANOTHER STEP FORWARD: THREE INNOVATIONS

To build higher and heavier other solutions had to be found. Eiffel's compressed air caisson without a decompression chamber had been quite rightly condemned by the health regulations of a more exacting Labour Inspectorate. Three decisive innovations came to the rescue: incisions, injections and inclusions.

Incisions

To dig a pit is to make an incision in the ground and this raises two particular problems: the extraction of the excavated material and the consolidation of the walls. These operations are quickly thwarted by the presence of water yet twentieth century needs have called for incisions well below the level of the water table.

Oddly enough, it was clays saturated with water, notorious for their instability in the excavation of cuttings, which provided a solution to the problem. A technique was developed whereby the particles forming the clay were dispersed in a colloidal solution to constitute a kind of mud. This mud was then given a density greater than that of water by the adjunction of heavy mineral particles. This heavy, viscous substance, fed into the incision through a central tube, cleaned out the hole by causing all the debris or cuttings produced by the boring implements to float to the top; at the same time it prevented the walls of the hole from bulging inwards or collapsing by forming what is known as a 'cake'.

Though the basic principles of the technique have not changed, it has had to be adapted to the different kinds of soil encountered. When drilling through clayey formations, for example, the cuttings tend to mix with the mud, contaminating it and lowering its viscosity; on the other hand, limestone and dolomite sands overload the mud with chemically inert substances. In either case, the mud then has to be purified and regenerated, an operation that is carried out in installations on the surface. If the formations traversed are too permeable, the water in the mud is drawn towards them, leaving behind on the walls of the hole a 'cake' that is too thick; the hole then has to be re-bored and the viscosity of the mud increased, etc.

This technique, which is continually being improved by oil-drilling engineers, has produced remarkable results, at least in the case of circular holes with a small diameter and up to 3000 m deep. Towards the 1950's it had a spin-off for civil engineering. For foundations, the principle of employing piles to transmit the load of a construction to a deeper and firmer substratum had been known from antiquity, but the operation was often thwarted by isolated solid objects, such as rocks or tree-trunks, which were difficult to force aside. The technique of the oil-drillers suggested the idea of casting the pile in the ground: a hole is drilled and then the drilling mud is forced to rise to the surface by pouring concrete down a central tube which reaches almost to the bottom of the hole. The soil acts as a mould and the resulting concrete pile frees the architect from the problems of pile-driving in urban areas (noise, vibrations and so forth).

There was further development towards 1960. Still with the aid of drilling mud, the idea was to cut rectangular incisions so as to be able to cast concrete strips in the soil to form moulded walls with a considerable bearing capacity.

Injections

On the esplanade at Ur the Sumerians, before constructing their homes, installed a number of vertical drains the purpose of which was to pass drink-offerings to the gods of the underworld (Wooley 1933).

The purpose of the injections practised over the last two hundred years, the history of which is retraced by Glossop (1960 and 1961), is less mythical and easier to understand.

In the course of working on constructions involving two levels of water (such as locks) on sites composed chiefly of sand, it was observed that the water could seep round or under the structure and scour its foundations. This is in fact the only instance in which the Biblical injunction 'Thou shalt not build thy house upon sand' holds true.

This problem confronted Charles Bérigny at the Port of Dieppe in 1802: the lock had been undermined and he decided to fill in the resulting voids with puddle clay by directing it with a gravity head through tubes that penetrated the floor of the lock. In other words his idea was simply to fill in relatively large voids.

The second stage was the attempt to create an impenetrable barrier to the flow of water not only by plugging the cavities in fissured or karstic rocks, or even in old masonry, but by filling at the same time all the tiny voids to be found in soils. Lastly, the third stage was to strengthen the soil structure itself by injecting special grouts that set hard.

The difficulty of these problems was due to the morphology and distribution of the voids in the soil, both of which vary enormously as the nature of the soil ranges from coarse sand to the finest of clays, with coarse sand offering a smaller total volume of injectable pores than fine clays, which have many voids that are very tiny and difficult to penetrate. The injection process depends not on the number of voids but on their size and interconnection.

There were technical problems to start with since the first grouts injected proved unstable. A big advance was made when it was discovered that cement and clay could be combined by maintaining the grains of cement in suspension in a solution of clay throughout the process of penetration under pressure. It was quickly observed, however, that if the injection pressure was too high the grout could force its way through the soil without impregnating it in the mass, so the grout was injected a bit at a time under gradually increasing pressure.

In the last twenty years or so, the Japanese have taken a further step forward: instead of filling in voids, consolidating the soil structure, or strengthening the substratum, they start by fragmenting it. The technique is known as 'jet grouting' and is applicable to clays. An unlined hole is bored and water is injected at very high pressure (200 atmospheres – and even considerably more) to break up the clay surrounding the hole into very small flakes while cement is injected, through another borehole, at the same pressure. The result is a soil ten to twenty times as strong as the original clay. With this brutal treatment of our mother earth, we are getting further and further away from the teachings of the prophet Smohalla!

Inclusions

Whereas incisions and injections are inventions not much more than a hundred years old, we have already seen that inclusions date back to the Sumerians, though it is only fair to add that they have taken a wide variety of forms over the last few decades.

For many centuries, these inclusions were of straw, reeds, wooden stakes, logs and fascines.

In the twentieth century, after the First World War, engineers placed reinforcements shaped like a lyre in the filling behind the abutments of bridges in order to strengthen the contact between the abutment and the ground. A little later the use of inclusions developed rapidly as a result of the Italian 'pali radici'; steel rods inserted into deep narrow boseholes and fixed by injections of a sand and cement mortar. Then came reinforced earth, with fibres or wires of metal or synthetic materials, evenly spaced (woven or non-woven fibre mats, plastic lattices or perforated sheets of plastic) (Figure 14).

Before turning to the consequences of these three techniques for vertical and oblique architecture, it is worth noting that they have very varied applications and are often used in the preservation and restoration of the world's cultural heritage. One of many examples is the consolidation of the Acropolis, the hill on which the Parthenon, the Erechtheion and the Propylaia were built in the age of Pericles, when the Greek civilization was at its peak.

In 1930, the slopes being in danger of sliding or collapsing, Balanos constructed massive retaining walls which have hidden any further deterioration of the rock from view. A recent study (1977) rev-

Photo 29. Monte Carlo. The flanks of the hillside facing the sea have been cut away. One can see the nine stages in the excavation: the black dots on the concrete mask are the anchorage points for the long tie-rods.

Photo 30. Another view of the same wall. The tie-rods are so long that they extend under the buildings on the upper level.

ealed the existence of 22 areas that were in danger of landslides and required immediate consolidation: it emerged that the outcrop of limestone overlies marls and conglomerates which have been undermined by the action of the climate, the vegetation and rainwater gullying down the slopes. The parts of the rock affected were first of all treated at the surface: drainage pipes were inserted to draw off the water, cracks were cleaned and filled with cement mortar. This was followed by an in-depth treatment of the rock mass which involved drilling and injections and the inclusion of prestressed rods of stainless ribbed steel anchored to the rock. Wire netting was then fixed at the surface and covered with a skin of concrete.

The Acropolis is not the only time-honoured site whose slopes are in danger. Many other hills bearing the inspired creations of man and impregnated with history, are slowly collapsing. The ancient papal city of Orvieto in Umbria is a case in point: the Italian Government, with the patronage of the Council of Europe, has just launched a big campaign to finance restoration work that involves many of the techniques we have described. As we shall see, Clement VII, sensing that Rome was about to be sacked, commanded the digging in this ancient hill town of a deep and magnificent well to provide water in case of siege; today – irony of fate – this once so precious water has become a nuisance since it is escaping from the town's drainage system and endangering the stability of the mountain slopes.

However, with the swift expansion of cities on the coast and at the foot of mountains (Hong Kong, Monte Carlo, etc.) and the driving of motorways through mountain regions, a further step has been taken. Man amputates the slopes, concealing this outrage against nature with a mask of concrete and carries his sadism so far as to force long iron needles into the wounded body of the earth in order to hold the mask in place (Photo 29). And so the gentle slopes covered with greenery are replaced by dreary expanses of concrete reinforcement anchored by tie-rods extending beneath the buildings on the higher ground – whose owners exact a rent for the occupation of their subsoil!

BUILDING HIGHER

It is not at all surprising that the first skyscrapers

were erected in Manhattan, since the underlying rock is of an excellent quality. But building to similar heights on other infinitely less favourable terrains was only made possible by the techniques we have just described in association with other advances in the art of foundations.

In this respect, one of the most interesting examples of the transition from an architecture that hugs the ground to one that soars upward is provided by Mexico City. We have already mentioned the poor quality of the soil in that area in speaking of the Templo Mayor of the Aztecs. In that city, the average height of the buildings was exceptionally low at the beginning of this century, when it was decided to construct the tall and ponderous Theatre of the Fine Arts. The Theatre lost its footing and found itself two metres lower down than its architect had planned. Later, the earthquakes of 1957 and 1975 would give further warning of the dangers lying in wait for over-ambitious architects.

Photo 31. Mexico City – a city of contrasts. The Latin American Tower and, just behind it to the right, the Fine Arts Theatre.

With the population exceeding ten million the city had to break with the tradition of low buildings.

The first skyscraper to rise from that soft soil, constructed by L. Zeevaert in the 1950's, was the Latin American Tower. It was built right in the heart of Mexico City (Photo 31), after a special permit had been granted by President Miguel Aleman, on a site northwest of the island of Tlateloco, where the Aztec Emperor Montezuma had formerly established his zoo. On the top floor of the tower some of the remains of the five-hundred-year-old Aztec period, found at a depth of 13 metres, are now displayed. The tower, with a remarkable four storeys buried underground in the clay, is really a ship. However, since the Archimedean upward thrust only counterbalances half of the weight (12 500 tons), the ship had to be prevented from sinking by supporting its bottom on long piles which transmit the extra weight to a deep-down layer of sand. Zeevaert and Newman made the calculations before any anti-earthquake regulations were issued but the building has stood up bravely.

Other skyscrapers followed, some of them violating the new regulations that were now in force. The very severe earthquake of September 1985 left the tower unharmed – but we shall see what happened to some of the other high buildings.

OBLIQUE ARCHITECTURE

With the techniques we have been discussing, it is now possible for deep foundations firmly anchored in the soil to withstand overturning stresses without flinching.

The builders of large bridges were the first to benefit from the new possibilities. It was good-bye to the enormous wooden centres used to support masonry arches, and even the first big arches in reinforced concrete, while they were being built. The modern bridges in prestressed concrete are constructed on the cantilever principle, with each successive 'voussoir' fitted to the one before by means of cables (Figure 65); the overturning stresses exerted when the arms of each side of the pier are unequal are absorbed by a deep-rooted foundation.

Now architects explored the possibilities of oblique

construction by building giant overhangs or soaring structures overarching vast areas and transmitting enormous oblique thrusts to the ground.

The architecture of the Olympic Games complex in Montreal provides examples of both these trends.

Beside a stadium with 70000 seats the architect Roger Taillibert has designed a tower 168 m high that overhangs its base by 65 m (Photo 32) and rests on foundations that reach a depth of 45 m in places. This 'mast' contains 18 levels of technical and sporting facilities serviced by a funicular and two lifts and, at the top, a space for storing an 18000 square metre awning weighing 200 tons which serves to cover the oval stadium during the winter months before being folded into its recess by cables descending from the top of the mast.

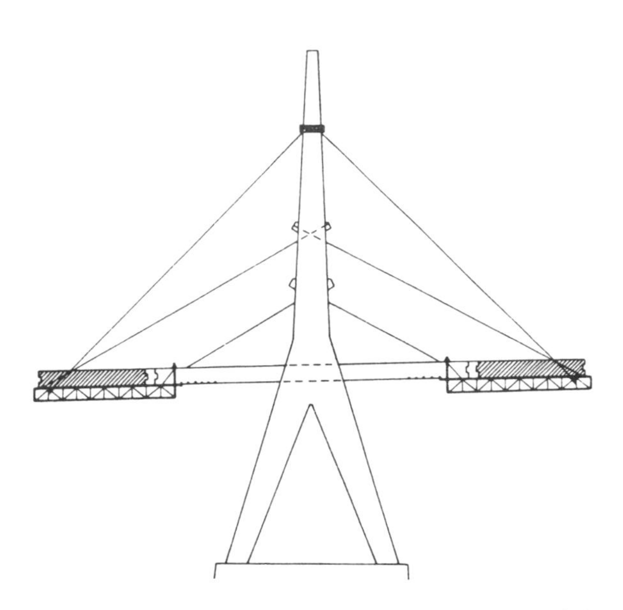

Figure 65. Overhang construction of the 250 m spans of the Lake Maracaibo bridge as early as 1960.

The mast is a tripod whose two front feet are obviously heavily loaded (50000 tons) while the back one has very little weight to bear and could therefore be vulnerable in the event of an earthquake. For this reason a number of cables have been anchored under tension at the bottom of very deep excavations to keep the back foot of the mast firmly in place. Construction has been twice interrupted but is now nearing completion.

The nearby velodrome (Photo 33) is covered by a gigantic vault with ribbing that recalls the veins of a maple leaf. This vault exerts heavy oblique pressures on the soil (Figure 66), which is composed of a very ancient limestone that is strong but contains microfissures, and even faults in places, in addition to voids of up to a metre wide. To absorb the thrusts involved, the rock was strengthened by means of injections.

The Romanesque or Gothic tradition was also to some extent an oblique architecture, but there is a

Photo 32. The Montreal Olympic Stadium with its oblique mast (Taillibert, architect): in the foreground is the velodrome.

Photo 33. Montreal, inside the velodrome (Taillibert, architect).

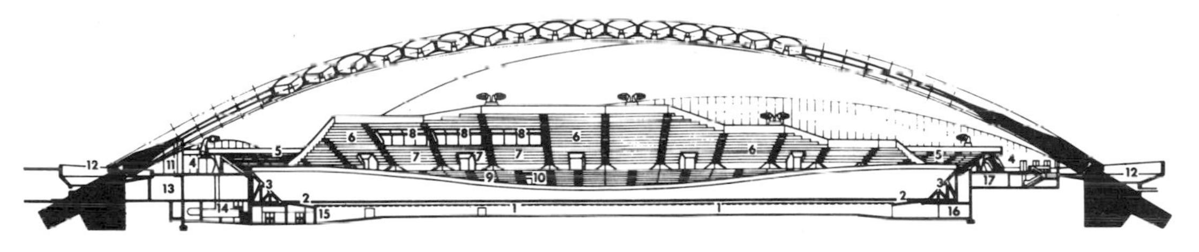

Figure 66. Montreal. Cross-section of the velodrome and its foundations.

world of difference between the Gothic flying buttresses and these modern ribs which channel enormous forces towards the ground! This leap forward was been made possible by the quality of present-day materials, especially recent types of concrete, and by the fact that geotechnical engineers can now treat the soil to enable it to bear very heavy oblique stresses.

* 11 *

The damming of rivers: Overloads to make the earth tremble

THE FORMS TAKEN IN AGRICULTURAL AND INDUSTRIAL CIVILIZATIONS

Whereas the traces of many monuments and palaces have disappeared forever it is relatively easy to find signs of ancient attempts to dam the flow of rivers in order to tame their excesses or, even more important, to bring fertility to arid lands. Despite the dangers and difficulties of such earthworks, a great many of them were built in the Mediterranean basin in Preroman and Roman times, as we can see from the map compiled by N. J. Schnitter (Figure 67).

The most ancient dam is the one at Jawa in Jordan (4000 B.C.), built of earth with a protective coat of masonry. The dam at Kafara, on Wadi Garawi, a tributary of the Nile about 30 km south of Cairo, dates back to 2600 B.C., i.e., just after the first pyramid of Saqqarah, which inaugurated the use of stone for construction in Egypt. It is approximately 12 m high and no less than 84 m broad at the base, that is, its breadth is seven times its height.

It was composed of two parallel embankments of rocks, laid without the addition of mortar, which form the upstream and downstream flanks of a filling of earth obtained from the surrounding hills mixed with materials from the bed of the wadi (Figure 68).

This dam makes it clear that the Egyptians had already arrived at the idea of a distinct separation between two functions – imperviousness (provided by the central filling) and stability (provided by the upstream and downstream dykes). This excellent approach, fundamental to the design of present-day dams (Figure 69), was, however, neglected in the nineteenth century for dykes constructed entirely of puddled clay which, though the slopes were relatively gentle, turned out to be unstable (see Figure 57): the water, remaining under pressure in the compacted clay, created internal stresses that provoked a lateral thrust when the reservoir was emptied.

Today, despite the intuitive skill of that Egyptian dam-builder, only the ruins of his work are to be seen opposite the plateau of the Lybian Desert on which the wonderfully preserved pyramids of Giza stand. All the other ancient dams are in ruins, too, moving vestiges of former agricultural civilizations that sought to escape from the harsh extremes of too little and too much rain; one day or another most of them were submerged and destroyed by a flood.

This danger was reduced when earth dams came to be strengthened with masonry. The masonry dam constructed by Nero for the nautical games at Subiaco, some 50 km east of Rome, lasted for almost thirteen centuries. It had the record height for the period of 40 m, but it was precisely this height which exposed it to a second danger – the undermining of its foundations by the infiltration of water under pressure. It was not until the present century that really big dams could be built, under which a dense network of injections was used to create a deep impermeable barrier to stop the underground seepage of water.

The ancient dams were simply a rectilinear and massive embankment to oppose the flow of the water. We call this a gravity dam. However, the idea of a curved dam with its abutments anchored in the sideslopes of the valley was clearly expressed in the sixth century in relation to a dam that Justinian (Emperor 527-565 A.D.) wanted to build at Dara, near the Turkey-Syria border: 'He did not build this dam in a straight line but in the form of a crescent curved against the flow of water so as the better to resist its violence' (Procopius of Caesarea, in 560 A.D.). Though the construction of this dam has not been confirmed, the idea is there.

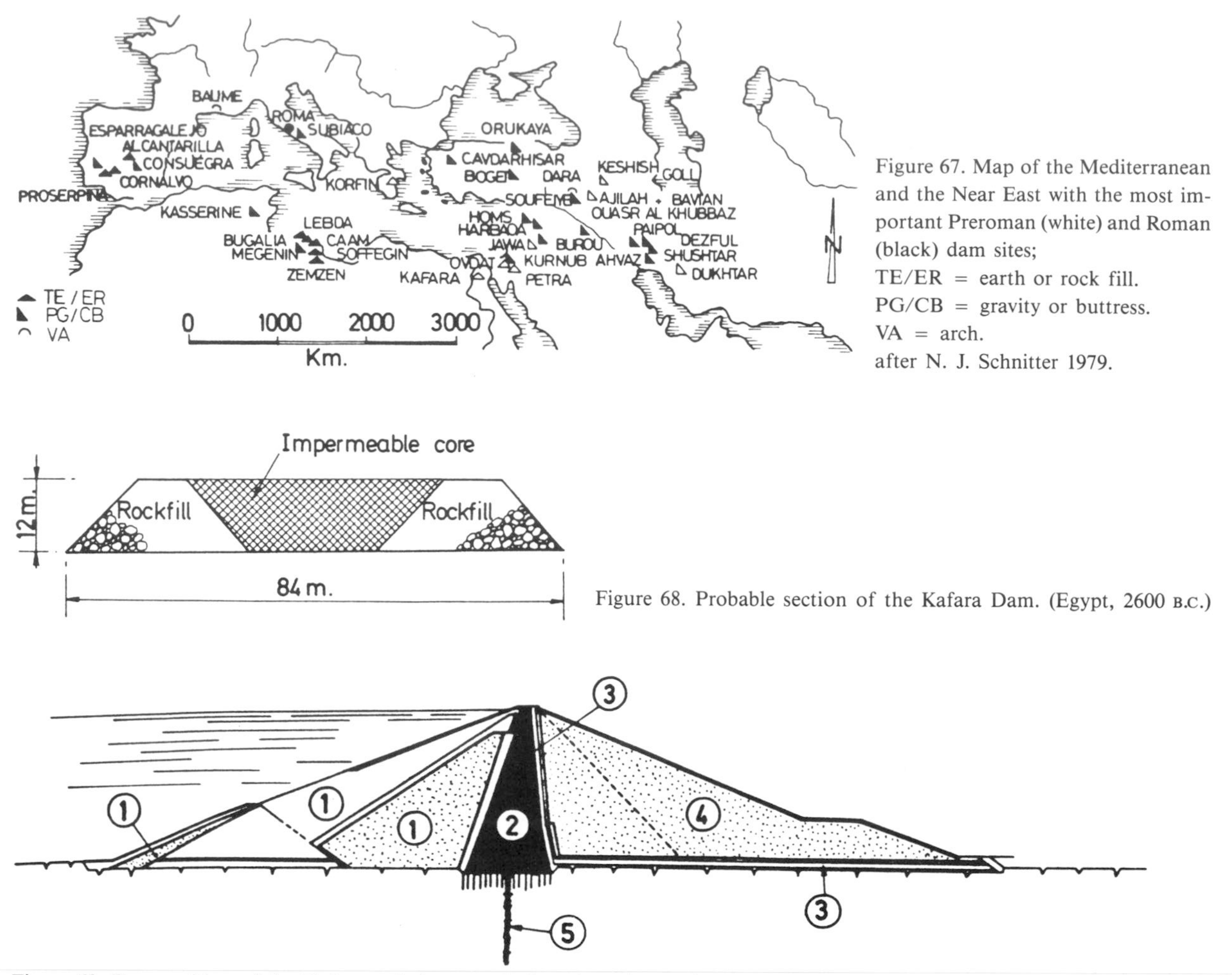

Figure 67. Map of the Mediterranean and the Near East with the most important Preroman (white) and Roman (black) dam sites;
TE/ER = earth or rock fill.
PG/CB = gravity or buttress.
VA = arch.
after N. J. Schnitter 1979.

Figure 68. Probable section of the Kafara Dam. (Egypt, 2600 B.C.)

Figure 69. Transposition of the Kafara technique: typical section of a modern earth dam.

1. Upstream shell: crushed rock, random fill, etc.
2. Clay core
3. Filter
4. Downstream shell: sand, gravel, crushed rock, etc.
5. Cast in situ wall or grout curtain.

Photo 34. Remains of the Kafara Dam: the upstream face of the upstream embankment, consisting of blocks of limestone laid step-wise (seen from the right bank). (Photo by J. Ph. Lauer)

However, with one or two exceptions, this approach was neglected until the twentieth century, when it was resuscitated with brilliant results. Now the enormous mass of the gravity dam would contrast with the light and elegant lines of the arch dam rooted in the flanks of the valley and in the bed of the river. More and more efficient concretes were developed and the thickness of these dams diminished. But earth is earth with all its obvious and less obvious peculiarities depending on familiar and less familiar geological processes. Soon there would be a divorce between the perfection of that manufactured material we call concrete and the imperfections of natural soils and rocks. It became a major art to reconcile the two. In 1959, the left-bank foundation of the elegant Malpasset Dam above Frejus in the South of France gave way and this disaster sounded the death knell for very high arch dams. The builder of Malpasset fell victim to his quest for a functional beauty – most unjust retribution! The ruins of this dam have the value of a symbol.

Photo 35. The Teton Dam (U.S.A.) after the collapse of 1976.

Ecologists will have to resign themselves to seeing the enormous volumes of gravity earth dams straddle the countryside. The thickness at the base of arch dams was only one seventh of their height. Man will now return to gravity earth dams with a base thickness five times their height – not quite the seven times decided upon at Kafara some six thousand years ago.

Admittedly, these 'artificial mountains' across river beds may not soar upward but they do at least give a feeling of security. However, even today and in the most technically advanced countries, albeit rarely, this security can be more apparent than real: the barrier of injections under the middle of the dam obstructs the infiltration of water, but if there is a single small 'window' left unblocked by the injections and if there happen to be erodable soils on the downstream side, the water of the reservoir will seep through this gap and undermine the dam.

This is what happened to the Teton Dam in the U.S.A. (Photo 35), even though it was only 90 m high and had been designed less than fifteen years before by a hitherto highly successful American team: the trench between the core and the barrier of injections was not impermeable everywhere and where it leaked the water seeped through and emerged on the downstream side on a rock used to found the dam but which was both permeable and erodable. Figure 70 shows the sequence of events suggested by Seed (1981) to explain the accident.

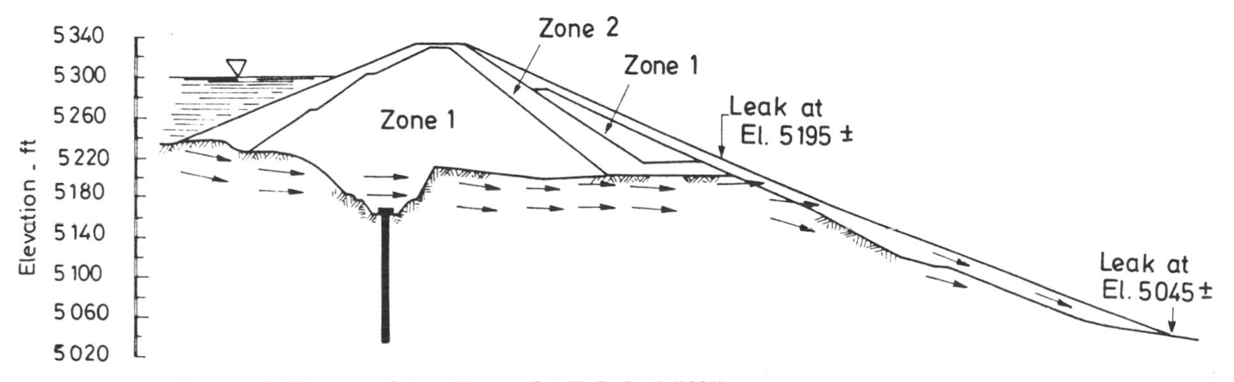

Figure 70. Teton Dam: probable cause of the collapse, after H. B. Seed (1981).

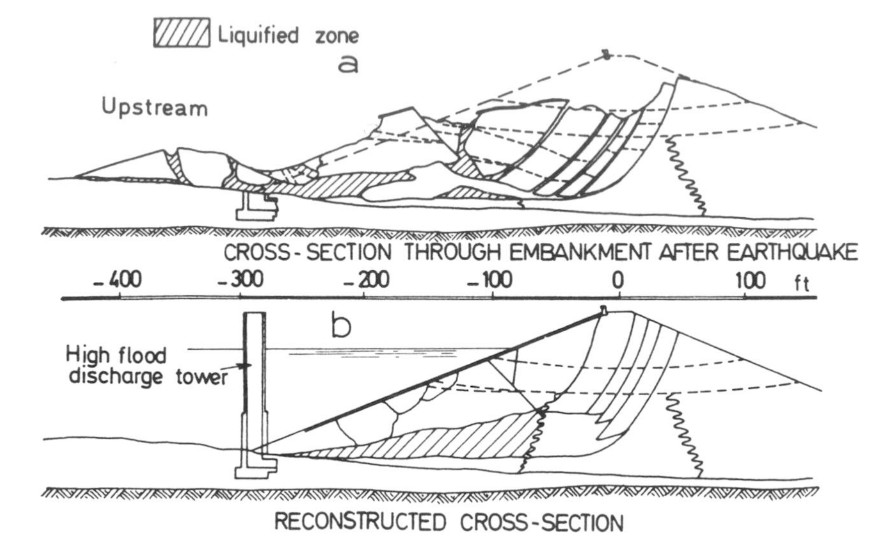

Figure 71. San Fernando Dam:
a) cross-section through embankment after earthquake
b) cross-section through slide area and reconstructed cross-section.
After H. B. Seed (1979).

THE EFFECT OF EARTHQUAKES: THE PARADOX OF ANCIENT AND MODERN DAMS

On 9 February 1971, at just after six o'clock in the morning, between 5 and 10 million Americans living in Southern California were woken up by the shock of the biggest earthquake to have struck the region in the past 65 years. In ten seconds it caused an immense amount of damage, including the collapse of the top ten metres of the San Fernando Dam, downstream of which dwelt some 80000 people. Luckily, the reservoir was not full – by chance, the water-level was just over ten metres below the crest (Figure 71). If it had been full there would have been the greatest disaster in the history of the United States (Seed 1979).

Only five years before, the stability of this dam in the event of an earthquake had been investigated by well-known experts, who had concluded that it was indeed stable whatever the magnitude of a possible earthquake.

The dam was located near the St. Andreas Fault, to which we shall return (Figure 122). The part that gave way was not particularly steep and the magnitude of the earthquake was only 6.6., but the acceleration of the horizontal movement of the ground was considerable: after nine seconds, it attained half that of gravity.

This near-disaster prompted a thorough investigation of the dams damaged by earthquakes in California and especially their reactions to the celebrated San Francisco earthquake of 1906.

There already existed a number of earth dams

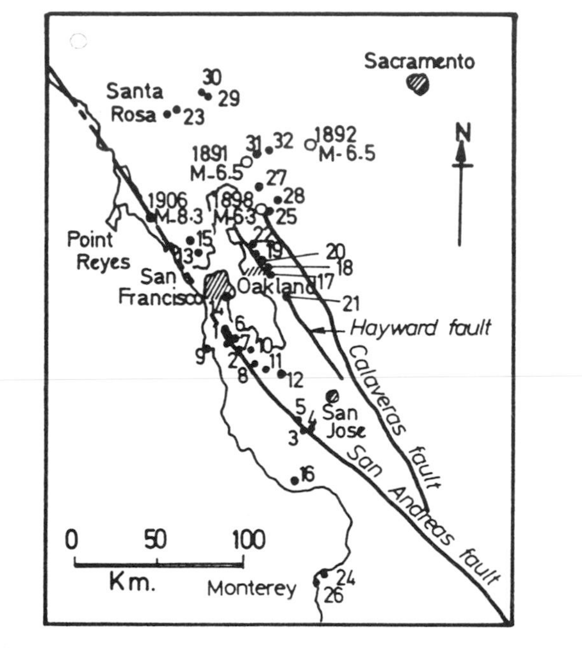

Figure 72. Locations of the 33 earth dams near the San Andreas Fault built before 1906. The San Fernando dam is much further south. After H. B. Seed (1979).

near the fault (Figure 72), 15 of them less than 25 km away, which had been subjected to that exceptional 8.5 magnitude earthquake for a full minute, and the nearest ones had suffered an acceleration at least equal to that measured in 1971.

It had to be admitted that the older dams, including many situated close to the fault, had stood the test magnificently. The feature they had in common was that they had all been constructed of clay on a clayey soil or on rock; in most cases the clay had been compacted by the most primitive means (by the passage of flocks of sheep, for instance).

In contrast, most of the modern dams that had suffered damage, and in particular the one at San Fernando, contained a sandy fill or sand saturated with water, or had been founded on a substratum of saturated sand. An explanation of the accident emerged: it turned out that as the shock waves of an earthquake pass through the mass of a dam they give up a part of their energy and that this lost energy increases the pore water pressure in the sands, provoking a kind of liquefaction capable of generating landslides (Figure 71). The clays of the older dams, on the other hand, had absorbed this energy without damage and without a dangerous rise in the pressure of the pore water, and the softness of the clay had attenuated the violence of the earthquakes. The accident to the San Fernando Dam became the starting-point for further progress in the dynamics of geotechnical engineering and in our understanding of the structure of saturated sands, some of which could liquefy under the influence of an earthquake. As one can see, older methods are not necessarily more vulnerable and the so-called modern techniques sometimes include hidden weaknesses.

INDUCED EARTHQUAKES: DAMS THAT MAKE THE EARTH SHUDDER

Dams may be sensitive to earthquakes, but they themselves may also, after their reservoirs have been filled with water, become the cause of earthquakes. A few spectacular incidents started off a lively discussion of the subject and the argument was subsequently fanned by the development of nature conservation associations.

It all began in 1962. The managers of the Rocky Mountain Arsenal near Denver, Colorado, wanting to get rid of a lot of polluted water, decided to inject it at very high pressure into a borehole 3700 m deep; in three years they pumped in 500000 m^3. A few weeks after the end of the pumping, sudden tremors were felt in the neighbourhood, and there were more in 1966 and 1967. In the same State, similar cases were observed near an almost exhausted oil-field around the periphery of which water had been injected at pressures of 250-280 atmospheres in order to revive the flow. Other earthquakes – some very severe – had been registered after the building of dams. So a table was drawn up concerning 19 dams filled since 1960 and the characteristics of the earthquakes that had occurred before, during and after the filling of the lakes.

Five of the earthquakes that had occurred after filling had exceeded a magnitude of 5.8. The one at Koyna (India) had killed 177 people and injured 2300, and the one at Kremasta (Greece) had caused a lot of damage (400 homes destroyed and 200 damaged). Public opinion was alerted.

The *New Scientist* of 11 July 1968 came out with the heading 'Fill a lake, start an earthquake': and in 1970, Unesco looked into the question and organized a meeting of specialists; the International Commission on Large Dams, having put the issue on the agenda of its 1979 Congress, conducted preparatory investigations which produced a more sober view of the matter.

The first problem was to discover whether the region in which the dam was located was prone to earthquakes and, if so, whether its seismicity had been measured over a long enough period and by reliable enough instruments sufficiently close to the flooded valley. These three conditions being rarely fulfilled, certain engineers tempted to shirk their responsibilities concluded that it was all a matter of pure coincidence and unfailingly pointed to the often lengthy time-lag between the filling of the reservoirs and the earthquakes. But the question could not be written off so easily: it had been noted, for example, that after the filling of the Kariba Dam (Zimbabwe), the epicentres of 159 earthquakes with a magnitude superior to 2 were located in the neighbourhood of the dam. Similar observations had been made by the Russians on the site of their Toktogul Dam over a period of 11 years before it had been built and these had been compared with the data recorded since it had been filled in 1972.

In brief, it was concluded that there existed a definite threshold beyond which earthquakes could be triggered off – a height of 90 to 100 m and a reservoir capacity of over 1000 million cubic metres. Beyond this threshold, the changes in local seismicity did not extend more than 10 to 15 km from the dam and were characterized by the following features:

1 Minor earthquakes are 2.5 to 4 times more numerous after the lake is filled;

2 The epicentres shift towards the region of the reservoir (more exactly, towards its extremities);

3 The focus of the earthquakes is not very deep (less than 15 km);

4 There exists a certain correlation between earthquake activity and variations of the volume of water in the reservoir, with a time-lag of two months.

But more severe earthquakes could occur, if the dam and its reservoir lie over a zone of the earth's crust that is being stretched horizontally – above a subducting plate, for instance (Gévin 1973) – or in a tectonically active zone containing faults and dislocations and where, in particular, water can seep down to a great depth.

* 12 *

Footholds on the sea floor

Well before the Christian era two Greek ports, Samos and Alexandria, had already left their mark on the sea floor near the coast. In his 'De Re Architectura', Vitruvius would later describe the highly ingenious methods imagined by the Greeks, and later by the Romans, for the construction of artificial islands protected from the sea and how they pumped out the water and fulfilled the ancient dream of descending beneath the surface in order to construct their moles on the ocean bed.

Chapter 12 of Book V is entitled 'De portubus et structuris in aqua faciendis' – 'On building ports and other structures in the water'. For the construction of moles, Vitruvius proposed alternative methods:

- gradual filling outwards from the shore;
- underwater construction within a wooden caisson, using stone, pozzolane and quicklime to drive out the water (Photo 36, lower scene);
- the construction of a double-walled coffer-dam. Between the two walls, reed sacks filled with clay are let down to the bottom and the inside of the coffer-dam is pumped dry with wheels and drums operating on the screw principle. These machines (Photo 36, upper scene) are bucket-wheels driven by men walking inside a big drum which also drives an Archimedean screw. 'If the drained soil is not firm enough', continues Vitruvius, 'piles should be hammered in and the foundations of the moles rested upon their heads, with channels left between the piles to serve as drains'.

The first and second methods were widely employed in antiquity while the third one is already a very advanced technique.

THE LIGHTHOUSE OF ALEXANDRIA

The ancients regarded the tower that Alexander decided to build on the Island of Pharos around 300 B.C. as one of the wonders of the world, and all subsequent towers erected off the coast to guide ships by means of lighted beacons took their name from that island. This lighthouse was still standing in the twelfth century, in the time of the Arab geographer al-Idrisi (1099-1164). The flame was known as the 'phanos' (from which the French word 'fanal', lantern, is derived) and al-Idrisi wrote that the light appeared to mariners 'like a star at night and a trail of smoke during the day'.

The tower was built on an island which Menelaus thus describes to Telemachus in the fourth Book of the Odyssey: 'In the midst of the waves lies an island that is called Pharos and is as far off the coast of Egypt as a boat borne by a following wind can normally travel in a day'. Homer thus situates Pharos a long way out from the mouth of the Nile. In Pliny's time the island was much closer to the coast and his pertinent remarks draw attention to the slow but steady reshaping of coasts and deltas: 'a large part of Egypt is formed of earth deposited by the Nile since Homer's time for that poet mentions the Island of Pharos as a day and a night's journey from the mainland'. Since then the evolution of the coast has gathered pace and Pharos is now incorporated in the sea walls of the port of Alexandria.

When the architect Sostratus of Cnidus built the tower at a distance from the coast, the island was open to the waves and so, to fix it more firmly to the rock, he shaped it like the trunk of a palm-tree, with a very heavy base. 'In addition to it being constructed of excellent stone of the kind known as Kedan', writes al-Idrisi, 'the blocks were cemented to each other with molten lead and the joints are so strong that the tower forms an indissoluble mass even though its northern side is continually beaten by the waves'.

Later, a mole some 900 paces long and a bridge were built, which carried a narrow track from Alexandria to the Island of Pharos. In the Pharsalia,

Photo 36. Illustration for Vitruvius' treatise, engraved by Perrault in 1684 for the Paris Royal Academy of Science. The upper scene shows how the seabed near the shore was pumped dry in Roman times.

Lucan tells of Caesar's arrival in Egypt: 'The zephyr swelled his sails and on the seventh night brought him in sight of the flames of Pharos lit up on the coast of Egypt; but the following day dimmed the nocturnal torch before Caesar entered the calm waters of the harbour'. According to Lucan, Sostratus of Cnidus, justifiably proud of his work, had his name engraved on the stones of the tower and then, having covered it with a layer of plaster, inscribed the name of the reigning monarch! His idea, it seems, was that the plaster would be as transitory as the glory of the king.

And he was right. A long time afterwards his engraved message appeared: 'Sostratus, son of Dexiphanes, from Cnidus, to the saviour gods'. Even if this is merely an anecdote invented by Lucan, it certainly has its place in a book about the traces of the past. It is not known at what period the lighthouse ceased to guide mariners but it is thought to have been destroyed by an earthquake at the beginning of the fourteenth century. Today, even the ruins have been scattered.

THE OSTIA LIGHTHOUSE

The Romans had a more difficult problem before them when Claudius, more than three centuries later, in about 42 A.D., gave orders for the creation of the great port of Ostia. Disregarding the advice of Vitruvius not to build a port at the mouth of a river, he told his architects to construct jetties beyond the mouth of the Tiber, the entrance to which was to be flanked by a lighthouse that would guide mariners in their approach to the new port of Rome. According to Suetonius (Claudius XX.3): 'Claudius built the port of Ostia by constructing two jetties which curved to the right and to the left and, in the deeper water, a mole to protect the entrance; to provide a firmer foundation for this mole, the ship which had brought the great obelisk from Egypt was first of all sunk in place; on top of this base a large number of pillars were built in order to support a very high tower that would, like the one at Alexandria, serve during the night to light the way for ships with its flames.'

Recent archaeological excavations on the site, now dry land, together with the description given by Pliny the Elder, make it possible to state that this flagship weighed 800 tons, was 104 m long, 20.30 wide and had six decks. After having had its decks dismantled, it was ballasted with concrete and sunk in 6 m of water. The lighthouse was founded upon the prow (Photo 37) and then the line between the ship and the mainland was filled in, with four smaller boats being sunk to provide the foundations for a new quay. This port and its mole was long praised by Roman writers; in particular, Juvenal relates the return of his friend Catallus and the perils he had suffered from a storm: 'The storm

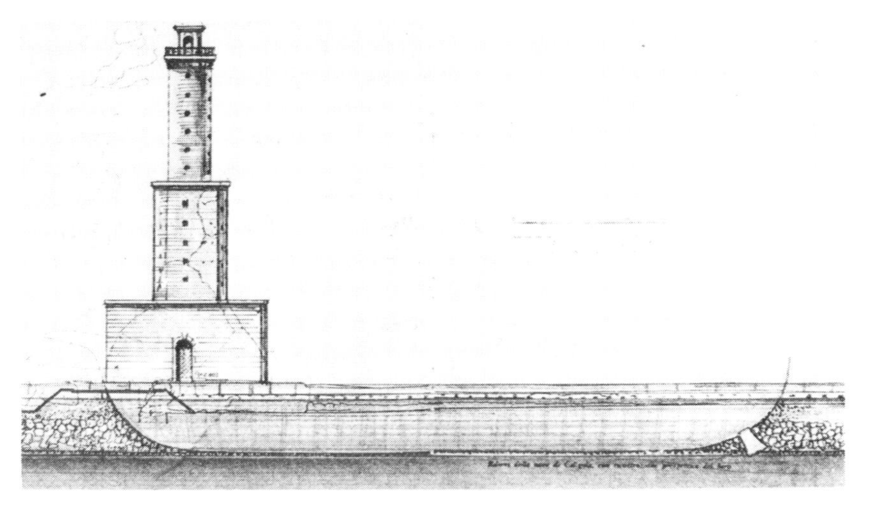

Photo 37. Reconstruction of the Ostia lighthouse and its foundations: the first off-shore platform. Latium archaelogicum. Perm. Ed. Il Turismo. Florence.

faded and hope rekindled with the sun. Then appeared the hills on which the town of Alba was founded. At last Catullus' ship passed the Tyrrhenian lighthouse and entered the port of Ostia, the moles of which extend beyond the lighthouse, closing off the distant waves, ... superb constructions more worthy of admiration than the ports hollowed out by nature'.

Let us spare a thought for that great ship which, after a life of combat, avoided the oblivion of burial at sea to become the pedestal for the Ostia lighthouse 'so that mariners could steer their course towards the nocturnal flames'. Nineteen centuries in advance, it foreshadowed the arrival on the ocean floor of the offshore drilling rigs and platforms.

THE EDDYSTONE LIGHTHOUSE

During this long hiatus of nineteen centuries, trade developed across the seas and with it the number of shipwrecks as vessels had their hulls stove in by rocks hidden just below the surface of the water.

Such was the frequent fate of boats which, at the end of the seventeenth century, set their course for the harbour of Plymouth. Some fourteen seamiles to the south lay the storm-beaten Eddystone Rocks, invisible at high tide. Despite the difficulty of landing even at low tide because of its jagged sea-washed shape, it was decided to build a lighthouse on the largest rock of the group. No less than four successive attempts were made to defy the sea: two lighthouses erected by Winstanley in 1698 and 1699, a third built by John Rudyerd in 1708 and a fourth by John Smeaton in 1759.

The first two, in masonry, were anchored to the rock by steel rods fixed in position with lead; the second lighthouse was really an enlargement of the first, which was considered too flimsy, in the form of an extra casing all around. On 27 November 1703 it was none the less destroyed by the sea, which carried away Winstanley, his assistants and the lighthouse keeper. The lighthouse was still too light but it was also insufficiently monolithic.

To avoid these weaknesses, Rudyerd cut some 'steps' into the pointed rock, on which he then erected a part wood part masonry structure that was anchored to the rock in the same way as before, with steel rods fixed in place with lead. The outer covering was of wood and, on 2 December 1755, the lantern set fire to everything made of that substance. The two keepers took refuge at the bottom of the external iron staircase, trying to avoid the falling drops of molten lead which flowed out of the red-hot joints of the upper storeys. The fire was so violent that it could be seen from Plymouth. One of the keepers died twelve days later, and seven ounces of lead were found in his stomach; the other was terror-stricken, fled and was never heard of again. After this tragedy, John Smeaton found the final solution by increasing the base area (Figure 73) and weight of the structure and vastly improving its monolithic qualities: dovetailed recesses were cut into the rock to receive stones specially shaped on the mainland to fit into them (Figure 74a).

A similar system was also used to interlock the stones of the upper courses (Figure 74b); at the same time, marble joggles were inserted into facing

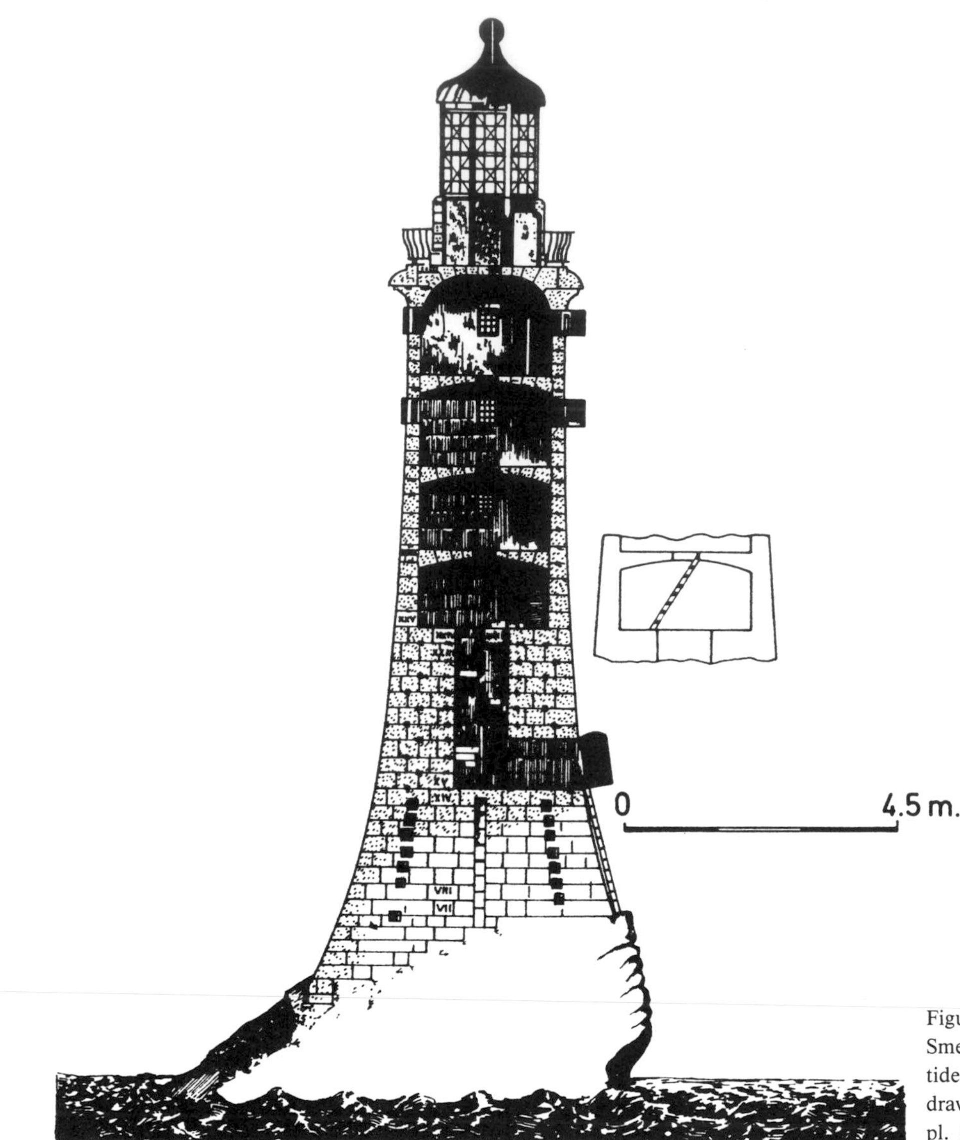

Figure 73. Eddystone. Section of Smeaton's lighthouse at low spring tide (engraving by E. Rooker from a drawing from Smeaton's 'Narrative', pl. 19).

holes between each succeeding course and all the joints were filled with a hydraulic lime mortar the exact composition of which had been specially studied by Smeaton.

The Eddystone lighthouse is a marvellous example of the stubborn determination and intelligence of its builders. Of particular note is the technique of prefabrication, which made it possible to interlock the stones in each course and cut down the time spent on the rock, and the development of new, sea-resistant mortars.

Smeaton's lighthouse lasted 120 years. The rock was so exposed and the structure so perfectly welded to it that the lighthouse would vibrate during storms. It was deliberately demolished in 1878 when Sir James Douglas built the fifth and last lighthouse on a more favourable rock some distance away, this time making use of a coffer-dam and pump in accordance with the third technique proposed by Vitruvius.

Nevertheless, the construction of British lighthouses in highly exposed places, in particular the one at Fastnet, would continue to be based on the ideas of Smeaton.

At about the same time, the Ar Men lighthouse was being constructed. It too was anchored to a rock in the open sea and had quite a long history.

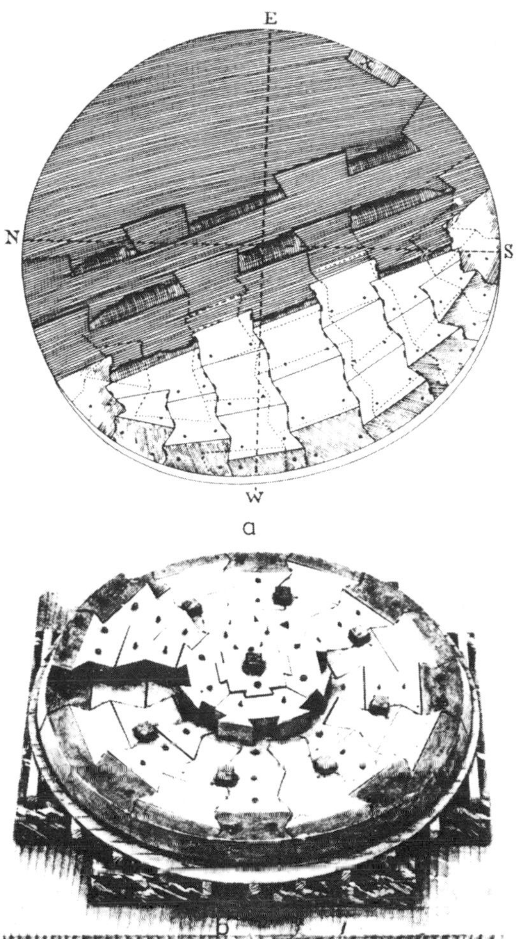

Figure 74. Eddystone. The foundations of Smeaton's lighthouse
a) Course IV. The stones were pre-cut on shore and then interlocked on the rock. The black dots represent the treenails which fixed the blocks to the rock.
From Smeaton's 'Narrative', pl. 10.
b) Model of course VII made by Jonas Jessop, 1757 (Royal Scottish Museum). The first complete course not in direct contact with the rock. The stones are dovetailed and marble joggles are used to unite successive courses.

Some fifteen kilometres to the west of the Pointe du Raz (Figure 75), a headland in Brittany, there is an area of rock around the Ile de Sein and a few isolated snags which ships sailing up from the Cape of Good Hope towards the North Sea have to circumvent to the west. It was most probably the perils of the Sein reefs that gave its name to a bay on the mainland just north of the Pointe du Raz – the Baie des Trépassés or 'Bay of the Dead'.

One of the most seaward rocks is known as Ar Men. It was on this rock, which just emerges from the water at half-tide, that it was decided to build a lighthouse. But whereas the Eddystone rock rose 4.50 m above the neap tide low water level, the Ar Men (Figure 76) one broke the surface by only 0.60 m and, what is more, the current when it did so was still 7 to 9 knots and changed direction after low water.

The first job was to provide an attachment to the rock. In 1867, the first year, it was only possible to land seven times for a total of 8 hours of work in the entire year. As the table shows, progress was slow:

Years	1867	68	69	70	71	72	73	74	75
Number of landings	7	16	24	8	12	13	6	18	23
Nr. of hours worked	8	18	42	18	22	34	15	60	110

The lighthouse came into operation in 1881 but, with its diameter of only 7.20 m for a height of 34.50 m, there were fears that it was too slender. People were familiar with the long history of the Eddystone lighthouse and with the disaster of the Minot's Ledge lighthouse in the U.S.A., which had been lost with all occupants during a storm in 1861. Investigations were conducted to find out whether Ar Men's base had been eroded by the sea: all the joints in Parker cement were scraped out and refilled with Portland cement, and it was eventually decided to add a protective jacket of masonry all round the tower up to a height of 11 metres. In 1923, in the middle of a storm, it was Ar Men's turn to catch fire and the interior was destroyed and had to be completely renewed.

It has been a curious combat fought against these underwater rocks, which prefer to rip ships' bottoms rather than allow themselves to be topped with towers. Today, in the twentieth century, the problem has taken a different form. Shipwrecks are now less of a danger than pollution and the oil-slicks which cover the water when giant tankers are breached. After the 'black tides' caused by the Amoco Cadiz in 1978 and the Tanio in 1980, the International Maritime Organization suggested shifting the international sea routes away from Brittany by establishing a lighthouse in deep water. This was a new departure in civil engineering: since the time of Eddystone and Ar Men, lighthouses have been constructed on the seabed but never at a depth of more than a few metres. The idea was to build one in water 127 m deep nearly 80 km to the southwest

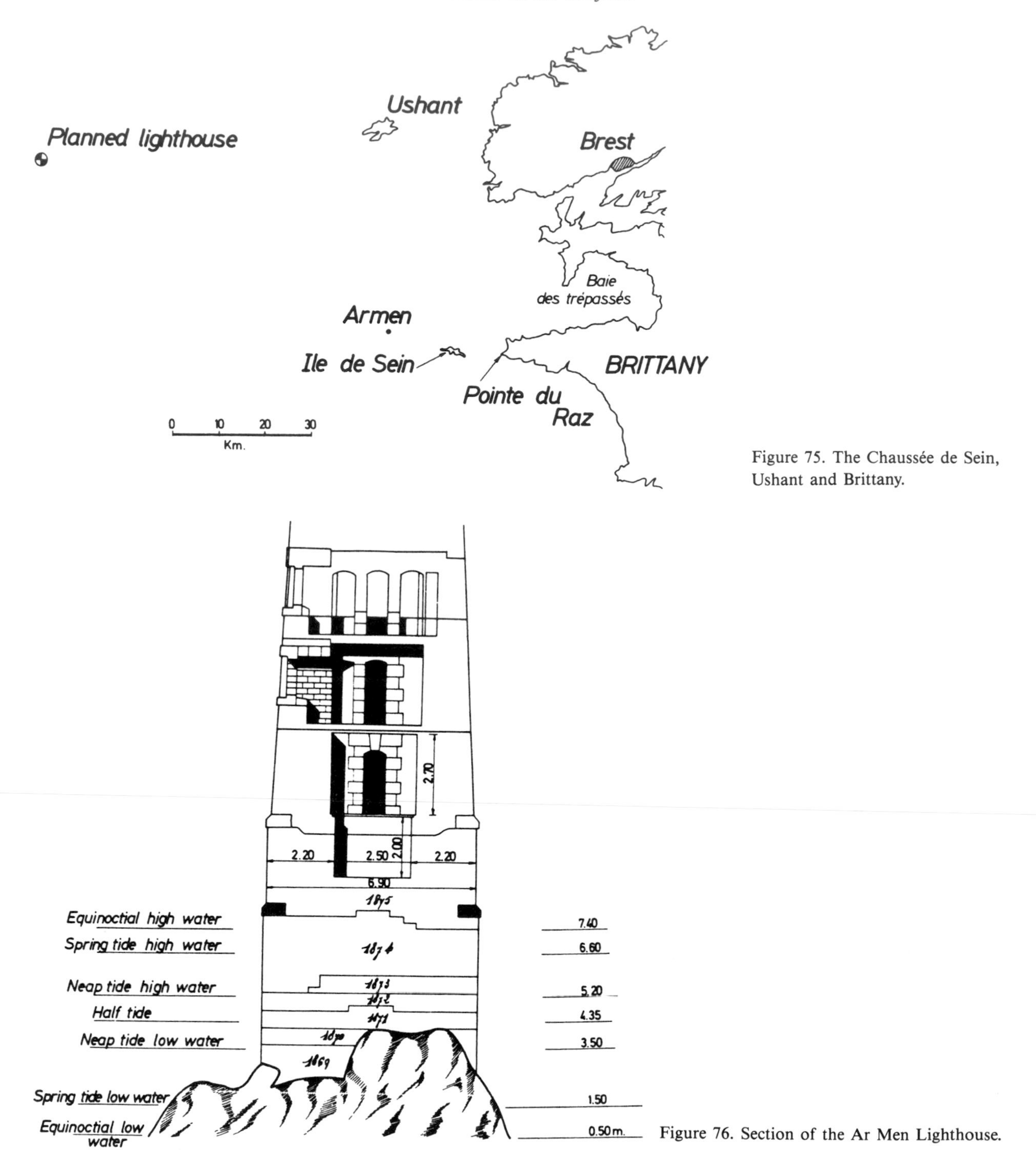

Figure 75. The Chaussée de Sein, Ushant and Brittany.

Figure 76. Section of the Ar Men Lighthouse.

of Ushant. The 25000 ton structure was to stand on three legs (Figure 77) which support a slender tower with a helicopter pad and a lantern controlled by radio from the mainland.

Here the foundation would be on a calm bottom well below the violence of breaking waves. But whereas battered reefs generally provide the lighthouse-builder with a hard rock for a base, the sea-floor much lower down is not always ready to bear such heavy loads safely or to resist the sliding and overturning forces exerted on a structure exposed to the waves and the swell of the open sea.

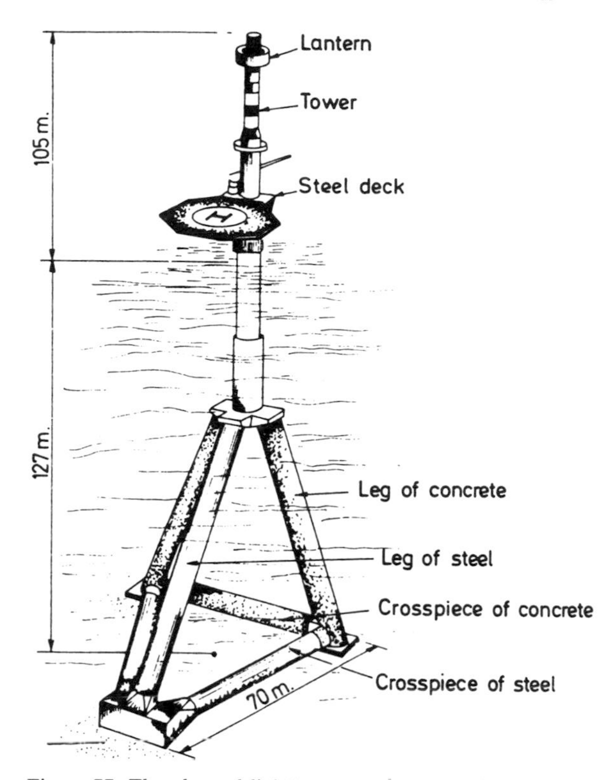

Figure 77. The planned lighthouse to the west of Ushant.

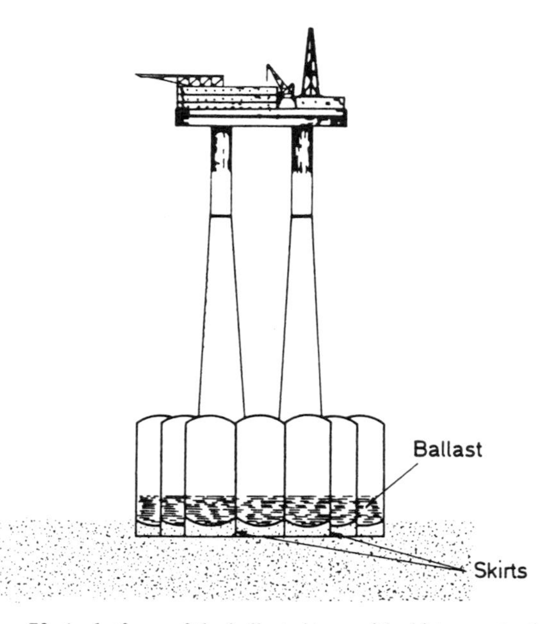

Figure 78. A platform of the ballasted type with skirts penetrating a few metres into the seabed.

In fact, this lighthouse has already run into the problem of what engineers call liquefiable sands, and the plan has now (June 1986) been abandoned.

OFFSHORE PLATFORMS

This leads us to the question of those other enormous loads now pressing down on the ocean floor – in the form of offshore platforms. Though oceans cover 71% of our planet's surface, the submerged portions of the continental masses that we call the continental shelf account for only 20% of this area – and it is here that 22% of the world's oil and 19% of its natural gas are to be found.

The seabed has already been required to support a platform weighing 900000 tons (Statfjord Field in the North Sea) and in 1989 the heaviest one of all, the Gullfaks C platform of 1400000 tons, will come into operation. But here, too, it seems that a rich oilfield is not necessarily compatible with the firmest of foundations – as if the seabed is unwilling to have its riches plundered.

Up until recently, the platforms in the North Sea rested quite comfortably on compact sands and stiff clays and their enormous weight forced their skirts no more than a few metres into the seabed (Figure 78). Tomorrow, the Gullfaks C platform will stand on soils the top 40 m of which are relatively weak sands and clays. To cope with this, the Norwegians propose to fit their platforms with very long skirts that will penetrate more than 20 m into the bottom of the sea. The difference in approach between offshore and dry land construction is immediately apparent: it is no longer a matter of building upwards from scratch even taking, as in the case of the Tower of Pisa, over two hundred years to finish the job, but of assembling the platform on the surface and then immediately weighing it down with enough ballast for its base to sink deep into the ocean floor (Figure 78).

However, the ocean floor sometimes gives the impression of balking at burdens that are too heavy: the seabed of the Ekofisk oilfield in the Norwegian sector of the North Sea, for example, has sunk 3 m since it has had to support the platforms. Admittedly, at Maracaibo or at Long Beach near Los Angeles, the pumping of oil has caused settlements of some 8 m at the surface but, in the case of the North Sea, specialists had ruled out this eventuality because the oil was being extracted from a reservoir rock. Ekofisk will remain in production up to the year 2010 and efforts are now being made to find a way of halting this settlement, which is bringing the decks carrying the technical equipment and living quarters for some 400 workers closer and closer to the surface of the sea. For example, Phillips Petroleum is thinking of injecting nitrogen to replace the oil and maintain the pressure in the subsoil or alternatively, of abandoning the lowest decks and rehousing the equipment some 6 m above the present level.

* 13 *

Under the earth

The subterranean world has always fascinated man. At first as a refuge: 'Lo, there was a great earthquake... and the kings of the earth, and the great men, and the rich men, and the chief captains, and the mighty men, and every bondsman, and free man, hid themselves in the dens and in the rocks of the mountains'. So wrote St. John the Divine in the 6th chapter of Revelations – and we shall see how recent experience has confirmed that such places are useful retreats during an earthquake. There too both martyrs and carbonari used to hide. Man has always felt drawn towards those wondrous caverns concealed in the body of the Earth, attracted by the unknown, by the sense of danger and the mystery of those regions, peopled in myths and legends with sibyls and nymphs, which some religions see as the seat of eternal damnation and others as the realm of bliss, while to certain heroes of the Arthurian legends they were the place of repose.

In the following pages we shall go 'down into the earth' and speak of the invisible art that both now and in the past has made this adventure possible.

Archaeologists and ethnologists have made the caveman into the archetypal occupant of the subterranean world. Yet, several thousand years before our era, it was already the theatre of military, industrial and agricultural activities.

ANCIENT MILITARY ACTIVITIES

Undermining is one of the oldest strategies in warfare and often provided an occasion to display the utmost ingenuity. As late as 1683, the Turks used it in their attempt to capture Vienna, perhaps the last real siege of a walled city. Long before, Vitruvius (Book X, Chapter 22) relates, Tryphon, the architect of Alexandria, had given proof of his cunning.

'The town of Apollonia was also beseiged and the enemy dug a sap in order to enter the town without being seen. Having become aware of this plan, the besieged population was stricken with terror since they had no idea where or when the assailants would appear. This uncertainty made their courage falter until Tryphon, the architect of Alexandria, who was in their midst, had the idea of digging a number of countersaps extending an arrow's flight beyond the ramparts. In each of these tunnels he hung brazen vessels. It so happened that, in the tunnel closest to the one which the enemy were working, the vessels trembled with every blow of the pick they gave, thus divulging the place towards which the enemy sappers were tunnelling in order to break into the city. Once this place had been accurately located, Tryphon ordered great cauldrons of boiling water and molten pitch mixed with sand to be made ready just above where the enemy were tunnelling and, after several openings had been made in their mine during the night, he had all the contents of the cauldrons suddenly hurled down upon those working below so that they were killed'.

MINING

In the Middle East, the copper age began about 6000 years ago, the bronze age about 4500 years ago and the iron age about 3400 years ago. To obtain the ores for these metals it soon became necessary to construct underground mines. These were not the very first mines since men had dug shallow tunnels in their search for flints in neolithic times, using the bones of animals as tools (Figure 79b); the skeleton of the unlucky miner clutching one of these implements (Figure 79a) is vivid evidence that such tunnels sometimes collapsed.

At the beginning of the copper age, the tunnels

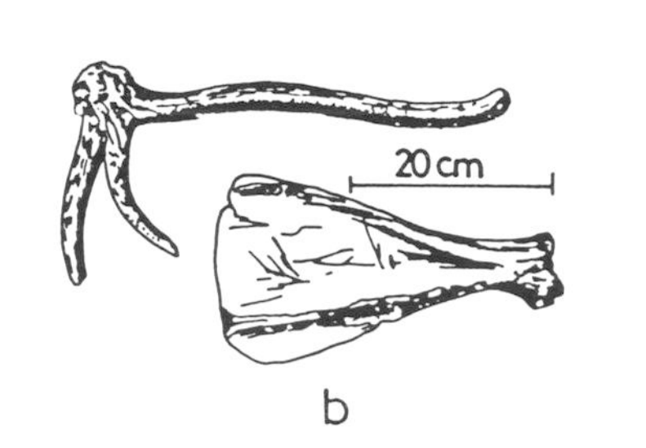

Figure 79. Neolithic flint mine (Obourg, Belgium)
a) One of the first victims of underground exploration: skeleton of a miner caught by a tunnel collapse while digging for flint with a scraping tool made from an animal bone. (Obourg, Belgium).
b) Two neolithic scraping tools: the antlers of a stag and the scapula of a cow, found near Worthing in England.

were usually dug into hillsides at a slight upward gradient so as to facilitate drainage. Later, the search for richer ores made it necessary to follow the slope of the rock strata and so deepen the shafts and lengthen the tunnels. The rock was often broken up by first heating and then suddenly cooling it with cold water. However, if we are to believe Figure 80, the methods of excavation and lighting used were rudimentary to say the least. The ore was removed from the mine by hand on wooden trays.

The lighting was provided by resin torches or animal skins soaked in fat, although Pliny (Natural History 33.70) states that the miners often worked in complete darkness. The Romans tried to introduce some form of ventilation but were not very successful. Strabo gives a description of the arsenic mines of Pinolisa at Pontus: 'The air in the mine was noxious; it was difficult to stand the smell of the mineral, and anyone working in that mine was promised an early death'.

The cavities formed by the excavation of the ore were separated by pillars left untouched 'to support the weight of the mountain' (Pliny, Natural History, 33.70). These pillars, called 'hormoi' by the

Figure 80. Greek miners. The ore, excavated with a pickaxe, was removed in baskets. Note the amphora used for lighting. (Corinthian clay tablet, 6th century B.C.).

Figure 81. Mining techniques, after Agricola (1556). Note the winches and the trolleys with four wheels.

Greeks, were cut away when the mine was abandoned, starting with the ones furthest from the entrance. Let us spare a thought for the miners who, in almost total darkness, had that job to perform!

In the Middle Ages, the most important work on mining and metallurgy was the De Re Metallica (1556) by Georg Bauer (1494-1555), who was more commonly known as Georgius Agricola. Figure 81 is a reproduction of an engraving from his book. It shows a mine with three shafts of dimensions 3 m by 1 m; the material was transported on small four-wheeled trolleys and the excavations were kept dry by means of hydraulic pumps. There were also shafts for ventilation or for escape in an emergency that were surmounted by a chimney shaped like a truncated cone with a winch on top. The circulation of air in the mine was sometimes improved by lighting a fire at the base of the shaft.

FARMING AND IRRIGATION

It is not our intention to relate the long history of the Persian 'qanats', Greek siphons or Roman aqueducts but simply to note that although the purpose of these underground constructions was to obtain or transport water, they were not situated within the water table itself. They were built in dry conditions with a protective barrier to keep out the water. This barrier was demolished at the last moment. Tacitus, for example tells how Narcissus had the plug of rock left to protect the tunnel dug to drain Lake Celano removed before the gaze of Claudius and Agrippina (Annals, XII, 56 and 57).

The vertical shafts of wells did not penetrate very far into the watertable either, but their depth and diameter were gradually increased as time went by. The Orvieto well is still one of the most interesting: 62 m deep and 13.40 m in diameter, it was commissioned by Pope Clement VII at a time when the Papacy was preparing for exile. The sides were formed of brickwork hollowed out to make room for two spiral staircases up and down which the local inhabitants and their animals went to quench their thirst. The beginning of a long journey towards the centre of the earth... Nowadays, downward exploration is more often via a narrow borehole than a wide shaft: as we shall see later on, the Kola Well has so far descended 12000 m with a diameter 50 times smaller than the one at Orvieto.

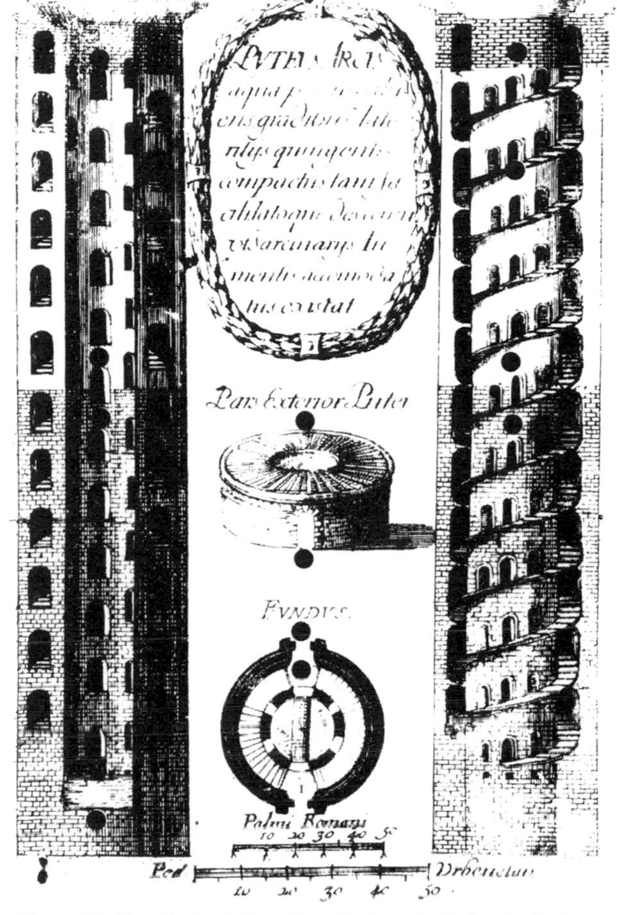

Figure 82. The shaft of the well at Orvieto, built by the Italian military engineer San Gallo 1527-1537. From an ancient drawing giving measurements in 'Palmi Romani' – 'Roman Spans'.

TUNNELLING UNDER WATER: THE TWO BRUNELS

The first big tunnel under a river dates back no further than the last century: it was the Rotherhithe Tunnel under the Thames in London, completed in the face of great dangers by Marc Brunel. Born at Hacqueville in Normandy, second son of a rich farmer who wanted him to be a priest, he had displayed royalist sentiments while at Rouen during the Reign of Terror. Forced to flee into exile in England, he eventually became one of the nineteenth century's most famous inventors and engineers. In 1818, he patented his 'shield', a device to facilitate the construction of tunnels. It was a sort of honeycomb frame, exactly fitting the cross-section of the tunnel being dug, which was pushed forward by

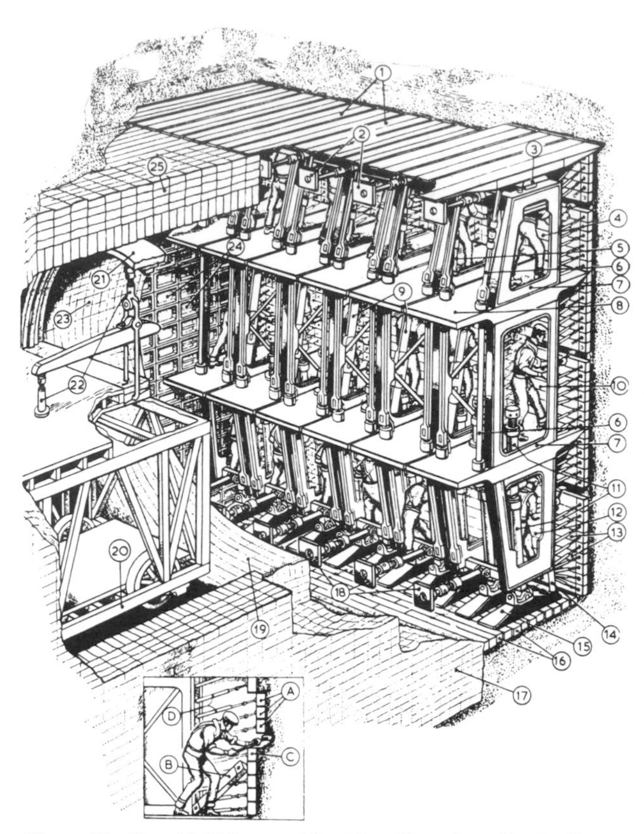

Figure 83. The shield invented by Marc Brunel to dig the first under-river tunnel within the London clay.

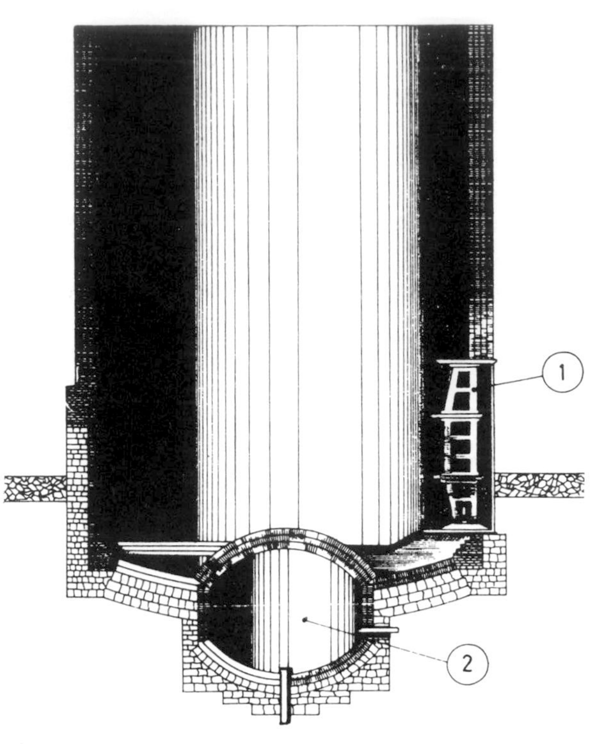

Figure 84. The shaft at Rotherhithe from which Marc Brunel advanced his shield beneath the Thames (1825).
1. The shield ready to advance laterally
2. Drainage sump.

horizontal jacks bearing against the newly completed brickwork just behind.

The workmen manned the cells of the 'honeycomb'. They would remove one of the poling boards at a time for just long enough to excavate to a depth of about a dozen centimetres behind it. When the soil behind all the boards had been removed the shield was jacked forward (Figure 83). In 1824, it was decided to utilize this shield to dig a tunnel under the Thames; geological reports on the subsoil had asserted that the shield could advance within a layer of clay with the crown of the vault no more than 4.20 m beneath the bed of the river. The work began in 1825 with the sinking, on the South Bank, of a large vertical shaft 15 m in diameter, at the bottom of which the shield was assembled (Figure 84). Then, on 28 November of that year, the hazardous struggle between man and water began. By May 1827, some 90 m had been excavated and the tunnel had reached the middle of the river; confidence was high and the Tunnel Company was replenishing its coffers by admitting sightseers at a shilling a head to view the daring enterprise. But at high water on 18 May 1827 the Thames breached its bed and poured into the tunnel: the geological reports were wrong, for the riverbed had been disturbed and instead of being homogeneous, the clay had been mixed with extremely permeable patches of gravel.

Despite the fact that the Rotherhithe Curate described the accident as 'but a just judgment upon the presumptuous aspiration of mortal men', Marc's intrepid son, Isambard Brunel, whose long career was as brilliant as that of his father, went down in a diving-bell to inspect the hole in the riverbed. He decided to fill it in with over a thousand tons of clay, which was put into sacks and lowered into position. The gamble worked, the tunnel was pumped dry and by November the shield had been repaired. Brunel organized a banquet in the tunnel to rekindle confidence in the enterprise, and the digging was resumed. But on 28 January 1828 the river invaded once again, causing a general panic among the workers. This time over four thousand tons of clay were deposited on the bed of the river. By now, however, the Tunnel Company had run out of money and it was not until 1835, by which time the two Brunels had proved their skill in a number of other ventures, that they decided to carry on. Even though they took the precaution of depositing a layer of clay on the bed of the river ahead of the shield, the Thames would pierce the roof of the tunnel no less than five more times before the job was completed. The air in the tunnel stank of sewage gas – men fainted and small flames flickered around the lamps. It was only in 1841 that Marc Brunel, now an old man, took his three-year-old grandchild by the hand and became the first person to cross under the Thames.

So much courage and stubborn determination excite the imagination. Brunel's achievement would stimulate many other projects – and even novels – in different parts of the world.

Was it, for instance, the heroic account of this tunnel that inspired Jules Verne, or was he thinking of the marvellous descriptions to be found in the book by J. Gaffard, published way back in 1654 under the promising title 'The Underground World, or an historical and physical description of the most beautiful caves and the Earth's most exceptional grottoes, and in general all the most famous caverns, cavities and underground passages and all the interesting things that are found therein'. Jules Verne's first book was in fact the Journey to the Centre of the Earth, which is little read today probably because the reading public, only too aware of the difficulties of steering an underground Nautilus through the magma beneath the earth's crust, finds it the one least easy to believe in.

Jules Verne's book is the story of an astonishing expedition into the bowels of the earth. Otto Lidenbrock, a German professor, and his companions trained themselves with 'vertigo lessons' in various craters and then penetrated that of an extinct volcano in Iceland. They crossed an immense underground sea on a raft of fossil wood. Later they entered a series of galleries: 'Sometimes a succession of arches appeared before us like the aisles of a Gothic cathedral; ... a mile further on, we would have to bend our heads under low arches of the romanesque type, resting on thick pillars half embedded in the walls'. The long voyage eventually comes to a happy end: on their raft, which 'rocked about on waves of lava in the midst of a rain of ashes and roaring flames', the Professor and his companions are ejected safe and sound from the volcano of Stromboli some 3000 kilometres away. A feast for the imagination!

Photo 38. Photo of an engraving by Riou from the French edition of 'Journey into the Centre of the Earth' by Jules Verne. 'The raft rocked on waves of lava'.

PRESENT-DAY REALITY: GREAT PROGRESS BUT CLEAR LIMITATIONS

Where have we got to in this conquest of the subterranean world?

There is no doubt that big strides have been made with the industrial era. The galleries of mines now wind their way several hundred metres below the surface; new caverns have been dug with spans that sometimes exceed 60 m; networks of bunkers several hundred metres deep in the earth have been constructed as the last refuge of a civilization; more recent years have seen the development of large shopping centres at the main junctions of busy underground transport systems; in the last decade or two underground reservoirs have been carved out to stock liquids and gases and the bits of drills prospecting for oil have bored thousands of metres down. As we shall see, the deepest borehole in the world has just attained a depth of 12000 metres.

Though the saps and countersaps of history may make us smile today, it does not seem that our colonization of the earth's crust can keep pace with the ambitions of our civilization. The last few years have revealed the limitations on underground urbanization. Since man's respiration pollutes the atmosphere in which he lives, the air of the subterranean world, a region which used to be regarded chiefly as a dwelling-place for the dead, has to be conditioned to make it fit for the living. This is why it will never become possible to put all road transport underground without a host of ventilating shafts dotting the landscape.

We shall never live in huge underground cities or drive through tunnels longer than thirty or forty miles. That dark expanse will continue to be received for non-polluting transportation, but the injections now being employed to render the soil firm and impervious will make it possible to pierce tunnels even under considerable depths of water.

The progress in engineering can be measured by the technical gap between the Rotherhithe Tunnel under the Thames and the Seikan Tunnel in Japan.

As early as 1945 it had been the ambition of the Japanese to link up two of their main islands, Honshu and Hokkaido, by means of a tunnel over 53 km long, known as the Seikan Tunnel. After 25 years of studies and reconnaissance, construction began in 1971 and the actual boring was completed in 1985. The first trains will circulate in 1988. At its deepest, the channel separating the two islands has a depth of 140 m and the tunnel passes 100 m further down. In other words it supports a pressure of over 24 atmospheres. Nearly a hundred direct soundings in the seabed had already revealed that the operation would be very tricky on account of the nature and variety of the terrain, portions of which were volcanic. Even so, the types of soil met with during the work, which took some fourteen years, did not turn out exactly as predicted. In particular, some unexpected geological faults encountered in volcanic terrain caused the tunnel to be flooded four times and, in the end, no less than ten such faults had to be traversed. When one is cutting a tunnel through a mountain one may release groundwater, but the flow gradually slackens with time and can be successfully pumped out. But a tunnel several hundred metres below the surface of the sea is quite a different matter: the intrusion of the water erodes the fissures, the flow increases and aggravates the situation. Altogether, the Seikan Tunnel cost 34 lives and injured over 700 workers.

Admittedly the Japanese were infinitely better armed than the Brunels and did not even begin the actual boring of the tunnel until they had injected the areas through which it was to pass. But a volcanic terrain with numerous geological faults requires very careful treatment. The Japanese ended up by piercing three tunnels instead of one: a pilot tunnel, a service tunnel and the twin-track tunnel originally planned. At the same time they found themselves obliged to develop a technique for exploring the soil far in advance of the trajectory, making horizontal soundings over 2000 m ahead of the part on which they were working. Altogether, they have invested as much as 2800 million dollars (three times the initial estimate) and it is by no means certain that the Japanese would undertake such a project today since the expansion of air traffic has cut the original traffic estimates from 17.2 to 2.1 million passengers.

A further stage in undersea tunnelling would be a passage under the Strait of Gibraltar. The author was asked to investigate the feasibility of such a tunnel on behalf of a major international organization. The best route, from Malabata (a few kilometres east of Tanger) to Punta Paloma (12 km northwest of Gibraltar), would take the crown of the tunnel's vault some 400 m below the sea surface at its lowest part, where the Strait is 300 m deep. A road tunnel would be out of the question because of the cost of ventilation, while a rail tunnel would have to be no less than 60 km long (since the shortest route would have to be much deeper) and then only on condition that the gradients at each end were constructed at 18 mm per metre, which is quite steep for trains and much steeper than for the Seikan Tunnel (12 mm per metre); at Gibraltar, the latter gradient would require a tunnel 74 km long. But the major problem is that the terrain to be traversed is extremely varied and we do not even know whether it would be a feasible – and economic – proposition to employ injections. It is known that there are geological faults but it is not known exactly where they are, and it is no use thinking they could be located by soundings from the sea surface since the total depth would exceed 400 metres. Nor do we know whether these faults are 'active' i.e. whether their edges are moving in relation to each other; this possibility cannot be ruled out since the African Plate is thrusting against Spain and slowly narrowing the Strait. In short, anyone building a tunnel would have to explore the ground ahead as he went along by taking horizontal soundings as at Seikan; he would therefore be unable to estimate the exact cost beforehand.

Indeed, this general problem is one that is faced by all underground projects. In his desire to go further and deeper, man is forced to feel his way forward since even the most sophisticated techniques of seismic prospecting from the surface can give no more than an imperfect picture of the situation.

Thus man has to dig his way beneath the land or the sea seeing little more than a mole, whereas the dangers lying in wait for him necessitate a much more detailed knowledge of the terrain he seeks to explore.

THE NEW CAVERNS

Towards 50000 BP, during the Middle Palaeolithic, the cave of Shanidar in Kurdistan was inhabited by Neanderthals; in 1950, the skeletons of men and women crushed by great blocks of stone which had fallen from the roof were discovered by Ralph Solecki. Though present-day caverns are much less dangerous for their occupants, they may sometimes represent a hidden threat to the neighbourhood. Today, huge cavities are being excavated deep underground, not as living quarters but chiefly for the purpose of storing oil and hydrocarbons. However, instead of doing this in large metal tanks hidden within these caverns, like those constructed during the Second World War, the rock is being left bare. Ecologists have welcomed the scrapping of all those containers that used to monopolize enormous areas in industrial regions but they are now expressing reservations about the locating of such underground reservoirs in soft and fissured rock, especially when they are used for storage at high pressure or low temperature – and their fears do not seem groundless since a number of these new caverns have leaked or collapsed. As a result, this technique is now subject to a number of economic and ecological restrictions, except in some countries of Northern Europe where the substratum is composed of hard rock.

DRILLING FURTHER AND FURTHER DOWN

Whereas the main handicap in driving tunnels under the sea is our inadequate knowledge of the peculiarities of the earth's crust, the biggest problems facing those trying to drill deeper and deeper towards the centre of the earth relate to pressure and temperature.

Deep down, the pressure in soft or plastic soils is such that the borehole closes up as soon as it is drilled, before it can be cased with steel tubes. Those drilling for oil are familiar with that critical depth of around 3000 metres beyond which many soils – and not just soft or plastic ones – show this tendency and they have had to develop special new techniques (drilling muds at very high pressures) to make further progress possible. Without this approach the Russians could not have succeeded in drilling the very recent Kola Well, at 12000 metres the world's deepest borehole, which lies in the Murmansk region of the USSR, north of the Artic circle (Kozlovsky 1984).

The earth's crust averages about 35 km thick in a radius for our planet of some 6000 km. Thus the Kola Well has so far explored only the upper third of it – rather less than that in actual fact since it is being drilled through the Baltic continental shield which despite recent erosion is somewhat thicker.

This vertical exploration has nevertheless revealed the existence of zones rich in minerals, confirmed the high pressures and temperatures deep down and revealed, at a depth of 9 km, an abrupt increase in the specific density of the rock and in the velocity of seismic waves. But whereas this abrupt increase had been expected to mark the transition from the upper granitic layer to the presumed basaltic basement in keeping with the 1926 hypothesis of Harold Jeffreys, the basalt has not been encountered.

Throughout its long journey down, the drill released copious flows of hot, highly mineralized water which had been trapped by the overlying layers of impervious rock. As this water escaped it

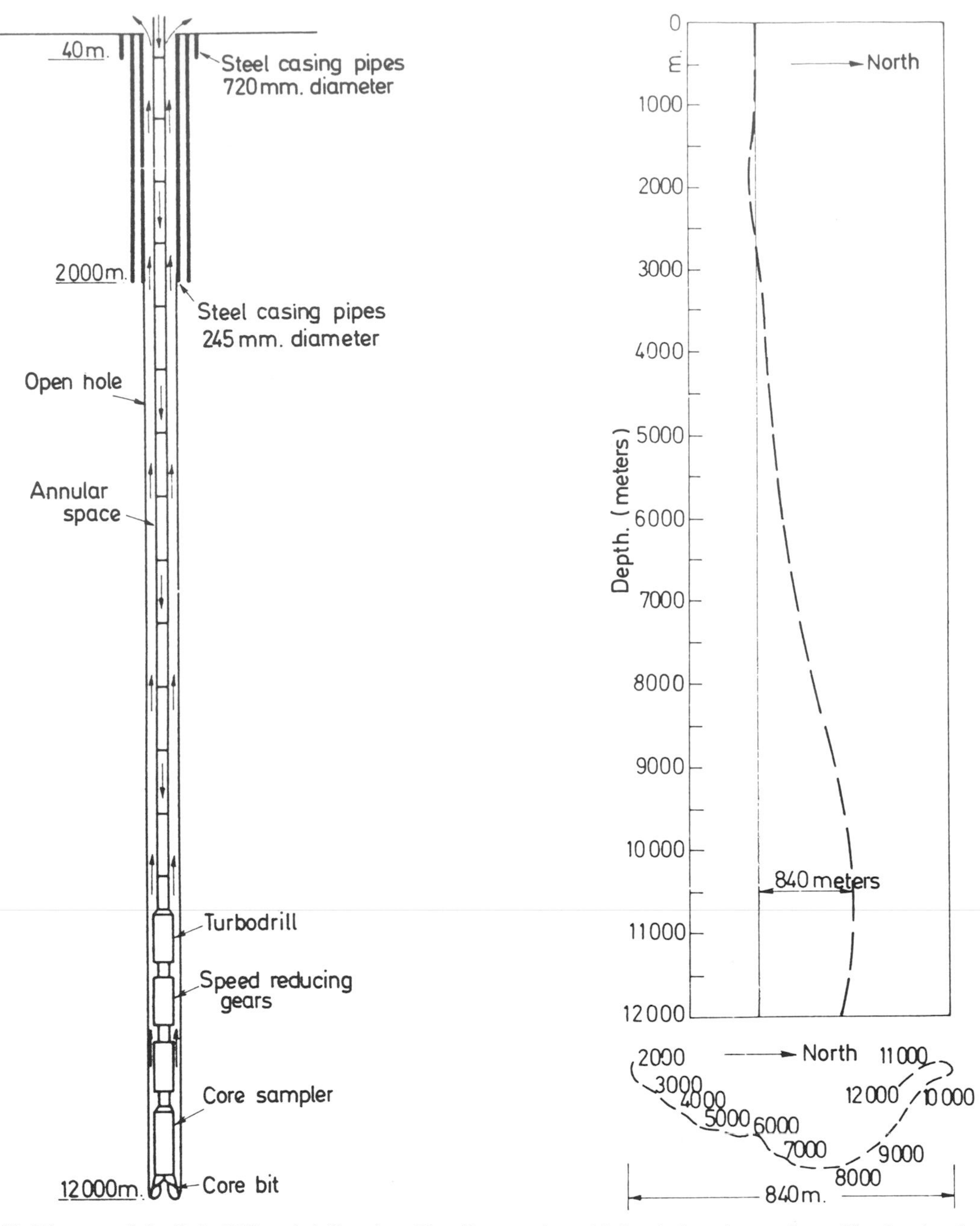

Figure 85. Diagram of the Kola Well and drill string. The diameter of the hole began at 720 mm, was then reduced to 325 mm and finally to 245 mm. (Kozlovsky).

Figure 86. Borehole trajectory in profile, showing the maximum horizontal wander of 840 metres. (Kozlovsky)

fractured the rock, which accounts for the irregular cross-section of the borehole beyond the depth (2 km) to which it is cased (Figure 87). Below 2 km the walls are prevented from caving in by the high pressure of the drilling mud, the circulation of which is accelerated by a turbine (Figure 85). The hole is not straight (Figure 86): the Kola Well has drifted from the vertical up to a maximum of 17 degrees at a depth of about 10 km, where the drill bit was 840 m to one side of the wellhead.

To overcome the high temperature and the corrosiveness of the hot mineralized water while at the

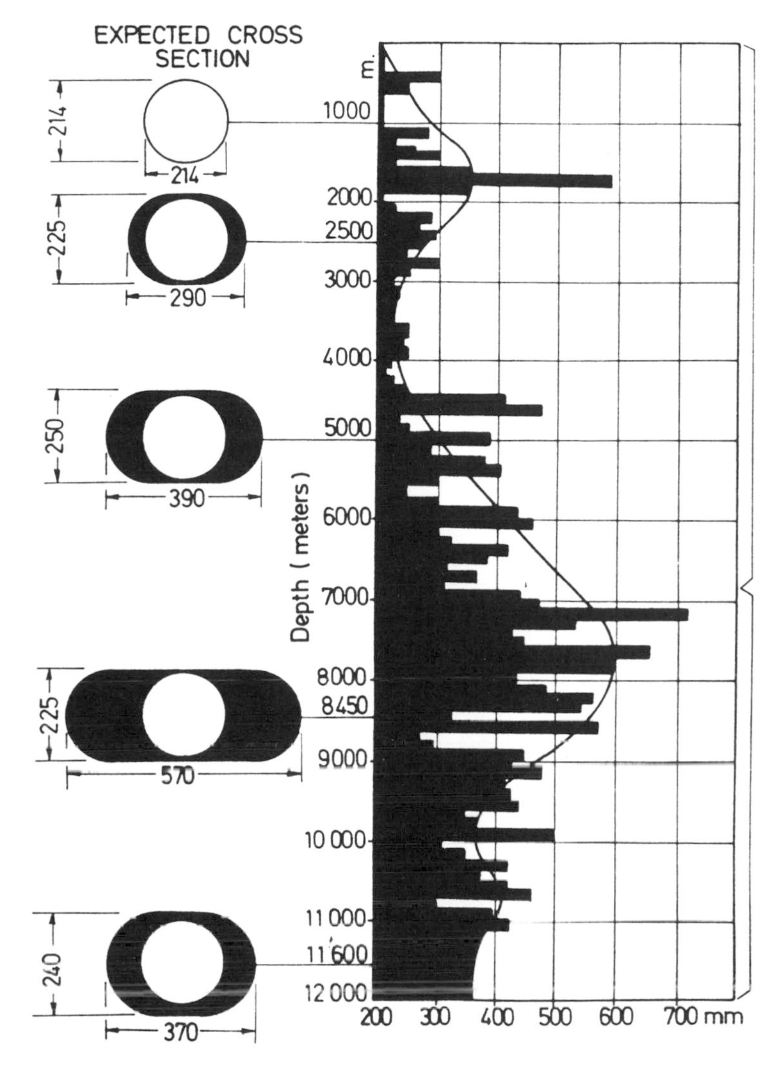

Figure 87. Cross-section of the bore-hole. The elliptical shape was due to unequal compressive forces in the rock or to unequal tensile resistance when trapped water was released (Kozlovsky).

same time lightening the drill string, the Russians have made use of special titanium alloys. They have now reached a zone in which the pressure of the rock exceeds 2400 atmospheres and say that they intend to go deeper still.

Kozlovsky speaks of the unsuspected wealth of minerals in the earth's crust, emphasizing that we have only just begun to scratch the surface. But this well gives some idea of the immense difficulties man will have to overcome in his drive to conquer the depths of the earth.

Part Three
The work of time

* 14 *

The permanence of the earth's burden

The earth has conserved only a few of the marks made upon it. The furrows made by the first ploughshares, the traces left by the first horseshoes and by the wheels of ancient carts have naturally disappeared forever. Nor is there now any sign of the saps and trenches dug out in the heat of long-forgotten wars.

But certain structures the earth has continued to support. How have they stood up to time? Has the earth's burden become easier to bear? Or has Atlas wearied of his task and shirked his responsibilities? Has he wilted quickly or slowly? And is man resigned to this or has he found some solutions?

The history of Chalco Amaquemecan contains the following proud statement by the founder of ancient Tenochtitlan: 'Nobody in the world will ever destroy or erase the glory, honour and name of Mexico Tenochtitlan'. The man made his boast – but we have already seen what became of the Huey Teocalli, the great Aztec temple in the centre of Tenochtitlan.

History is filled with this desire for the immortality of a work or person. Witness the inscription engraved by Darius on the head of a promontory at Behistun, some three hundred feet above the floor of the valley, recounting his triumphs some two thousand years ago: 'You who in the time to come contemplate this inscription engraved by my decree... do not erase or destroy it'.

Only great builders or conquerors can nourish the illusion that time will erase nothing.

THE EROSION OF TIME

Time has worn away the first monuments in sun-dried brick by transforming them into small mounds (tells) under which no doubt once glorious Mesopotamian cities still lie buried. The little rain that falls in the arid regions where the first ziggurats were built has gradually consumed them by seeping behind the dry crust formed during the preceding dry season and detaching it. A few still subsist, but one has only to glance at the plan for reconstructing the ziggurat of Aqar Quf, drawn up by some Turin architects (Figure 8), to see how much time has eaten into and shrunk this monument, one of the few of its kind that still survives.

Nor must it be thought that time has spared the sumptuous pyramids of stone of the 3rd and 4th dynasties.

The Pyramid of Cheops, the seventh wonder of the world, is scarred by the sandstorms which have eroded its once beautiful stonework (Photo 39). Here and there discreet attempts have been made to repair the time-worn blocks with concrete, coloured to look like stone. And what is to be done about the first of the great stone pyramids, the Pyramid of Saqqarah, which now has gaping holes and much more damaged ridges (Photo 40 and 41)?

It has often been said – and not without reason – that time is a thief.

THE ENEMY WITHIN

The forces of destruction often penetrate into the very heart of the monument and its foundations. This is what happened to Borobudur (Figure 88), the greatest temple erected to glorify the name of Buddha built in about 800 A.D. The vegetation had not only smothered it from view but, in forcing its roots into the stones, had loosened and broken them up, allowing the abundant rains to infiltrate and undermine the foundations.

This lofty stupa rises in a number of terraces which form the horizontal steps of an immense retaining wall (Figure 89) built to stiffen a hillock

Photo 39. The backing stones of the Pyramid of Cheops have not been spared from erosion by sandstorms.

Photo 40. Ageing of the Saqqarah Stepped Pyramid.

Photo 41. Undermining (seen from below) at the Saqqarah Pyramid.

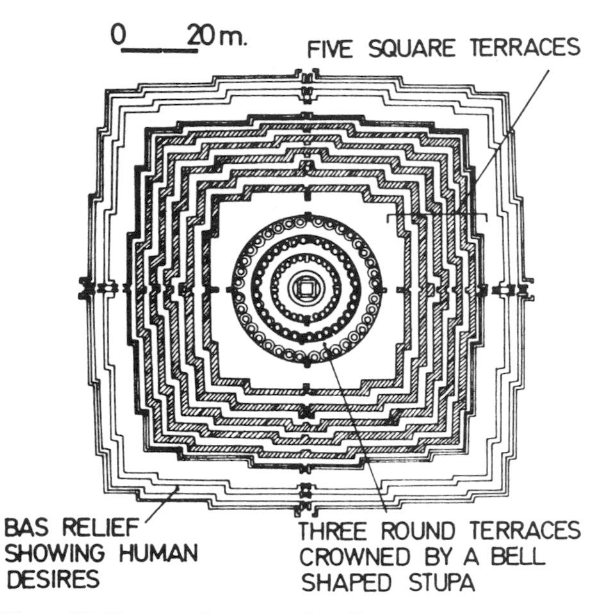

Figure 88. The central stupa and its five terraces.

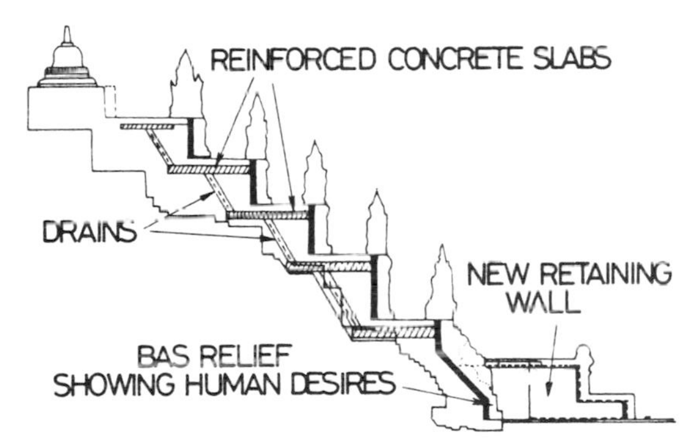

Figure 89. The latest restoration of the temple of Borobudur, by Unesco.

composed of tuf and clay. These terraces represent a symbolic journey: the pilgrims start from the sphere of Desires, to which man is a slave, then pass through the sphere of Forms in which man, having abandoned his desires, nevertheless conserves his form and his name, and finally reach the sphere of Formlessness in which, at last freed forever from all links with the world, he can taste of wisdom.

The 160 inscriptions and illustrative bas-reliefs at the foot of the stepped terraces, depicting the sphere of earthly Desires, are hidden from view by some 13000 m^3 of stonework. Was the intention to prevent the pilgrims from setting their eyes on these impure desires? Probably not. The most likely, and less moral, reason for this wall is that at an unknown period – possibly even during its construction – the temple threatened to slide down the slope and the only way to stabilize the monument was to quickly reinforce its base by adding this heavy covering of stone.

The fact is that the climate at Borobudur is hot and punctuated by the annual visit of the monsoon: between October and March the area receives no less than 2000 mm – about 80 inches – of rainfall, most of it concentrated in a series of extremely heavy cloudbursts. Great quantities of water accumulated in the filling on the tuf and clay of the hillock, under the stonework, and exerted pressure against the lower part of the wall, in particular the section representing the sphere of Desires.

After an initial and unsuccessful attempt at restoration at the beginning of the century, a new scheme to restore the temple, at a total cost of about $20 million, has just been completed with the technical and financial assistance of Unesco. This project has been one of the biggest of its kind ever attempted and has involved the cooperation of geotechnical and hydraulic engineers, architects, climatologists and so on. It is an impressive example of how teamwork between a number of disciplines can bring back to life one of the great monuments that comprise the heritage of mankind.

It was first thought possible to remove the strengthening wall from the foot of the stupa so as to reveal the sculptures depicting the sphere of Desires. But the idea had to be dropped, partly because of the cost and partly because of the fear of destabilizing the monument. The square terraces were dismantled stone by stone (Jayaputra, 1984) so that a layer of reinforced concrete could be installed underneath, with lead sheeting and pipes to drain off the water inside (Figure 89). The 240000 cut stones composing the outer surface of the temple were removed, numbered, cleaned, brushed and rid of organic matter with a chemical paste.

It is worth noting that the restorers made use of mortars composed of bituminous araldite and sand, the elasticity of which helps the temple to withstand the frequent earthquakes occurring in Java.

In less than twelve centuries the water of time, seeping little by little through the monument, had forced profound changes in the material expression of that journey through the stages of asceticism which the ninth century Buddhists had wanted to represent in geometrical form, and man has had to intervene twice in the present century to restore the shrine. How long will it now take, in a climate of such vigorous growth, for the new drains, so essential to maintaining the equilibrium of the structure, to become choked up?

EVEN THE MOST INVIOLABLE TOMBS UNSPARED

According to certain myths, children come from the depths of the earth: perhaps it is this vague memory of a pre-existence inside the earth-mother that accounts for man's powerful desire to be interred after death. 'Hic quo natus fuerat optans erat illo reverti' – Where he was born, there he desired to return. But is this tomb – this sanctuary – truly eternal? The great royal sepulchres were indeed built for eternity. From the Egyptian Middle Kingdom onwards it was the rock in the heart of a limestone cliff of the Libyan chain (Figure 90) that was chosen for the eternal resting-places of the Pharaohs. It is often thought that the only damaging violations have been at the hands of tomb-robbers. Not a bit of it. In 1975, some consultants from Brooklyn Museum found rapid deterioration on a massive scale: out of 62 tombs, 25 were half full of debris and 12 others had become inaccessible. A team of Egyptologists, architects, engineers and geologists brought together by the museum has recently demonstrated that the main cause of the deterioration of these tombs was the swelling, un-

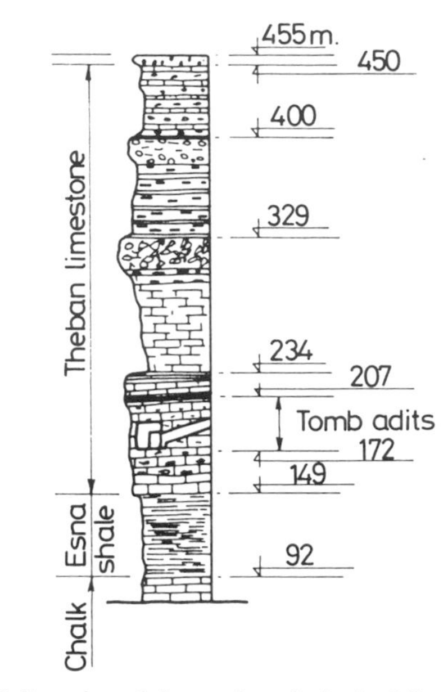

Figure 90. Location of the royal tombs in the Libyan hills near the Valley of the Kings.

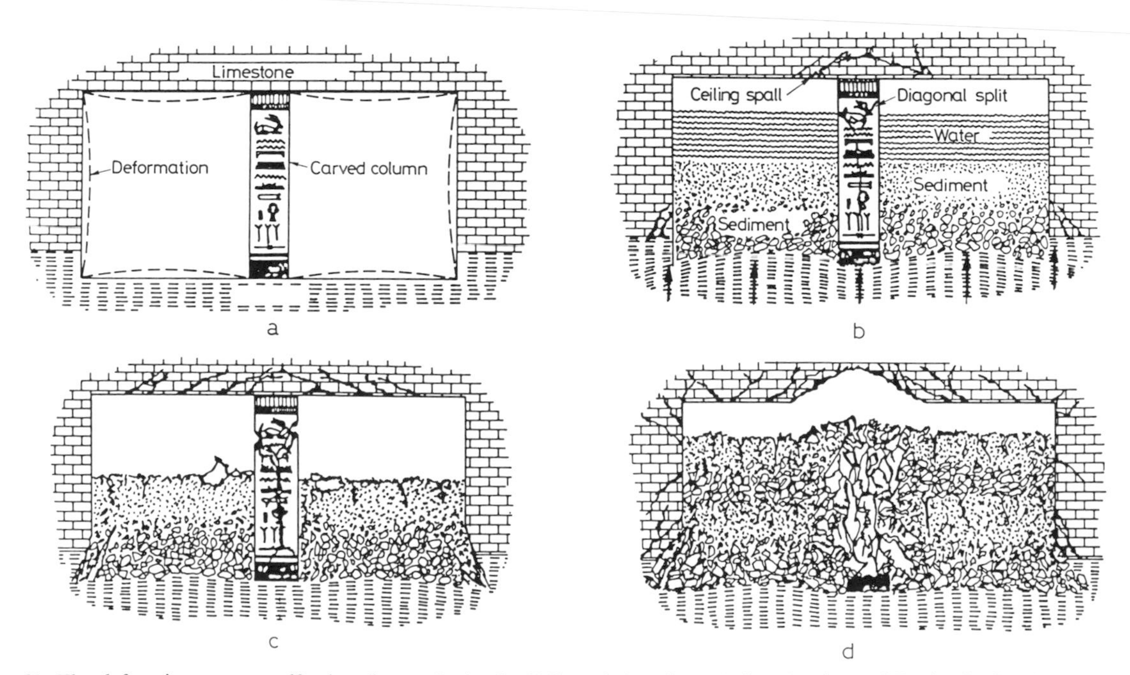

Figure 91. The deforming process affecting the tombs in the Valley of the Kings. (After Curtiss and Rutherford) (1981).

der the influence of rainstorms, of the clayey strata interbedded in the shale under the tombs (Figure 90).

The four diagrams provided by Curtiss and Rutherford (1981) (Figure 91) show clearly the process that has brought about the ruin of these burial chambers. The chambers had central columns which had gradually come to bear the entire weight of the rock overhead, a mass of limestone more than 150 metres thick. The columns (diagram (a)) had, with time, been crushed shorter and slightly cracked. Diagram (b) shows how, after storms, the swelling of the underlying shale exerted pressure from below and greatly increased the compression of the columns; the cracks widened and the columns punctured the roofs of the chambers (c). With the return of the dry weather, the shale diminished in volume, drawing the columns back downward. The roofs of the chambers followed, collapsed, and in doing so destroyed the columns (d).

One is nevertheless struck by the contrast between the solidity of these tombs, all built between 1550 and 1100 B.C. and the very brief spells of rain in a region for which the annual precipitation is only about a centimetre. Yet three thousand years of repeating this seemingly insignificant attack have sufficed to cause the ruin of these inviolable sepulchres. The astonishing thing is – how did this water penetrate as far as the burial chambers? The answer lies in the fact that the limestone around Thebes contains a great many vertical joints and patches of marl; by expanding slightly under the influence of the humidity, the marl eventually widened the joints and little by little made it possible after a long period of drought for the water to seep down.

Needless to say, modern man will try to conserve these tombs by diverting the rainwater on the surface and sealing the joints that open out in the chambers.

DESTRUCTION FROM BELOW

Around 2700 B.C., Mohenjo-daro was a modern city of the ancient world with some 30-35 000 inhabitants. It lay on the banks of the Indus and was the starting-point for an entire civilization which abruptly came to an end about a thousand years later. Archaeologists have recently discovered the ruins of this city, marvellously preserved over thousands of years under the alluvium of the plain. But no sooner had these ruins been uncovered than they were threatened by two enemies – water and saltpetre. The construction higher up the Indus of the Sukkur Dam has caused the ground water, which has a high concentration of salts, to rise and the sun, by capillarity, has drawn these salts up into the lower foundations of the city, swelling and splitting the brickwork at the base of the walls. But this is not all: the Indus itself is in the process of forming new meanders and is threatening to submerge the work of the archaeologists. Strange destiny for a civilization that disappeared and then re-emerged, only to return to oblivion once again unless something is done to save it. In 1974, when Unesco launched an appeal to international solidarity, the estimated cost of the initial programme of work already amounted to $7.5 million (Maheu, 1973).

DISAPPEARANCE OF THE LAND ITSELF

Having outlined a few of the slow-moving effects of time, against which it is possible for man to take action, we shall now look at another type of enemy, sometimes snail-paced in its progress and sometimes very quick, against which there is often little that man can do.

One example is the erosion of coastlines. Every year, stretches of history are swallowed up by the waves. Typical of this are those watchtowers, constructed by the Romans well behind the cliffs to guide mariners and which, many centuries later, are claimed by the sea. This is what happened to the Tour d'Ordre, which used to overlook the Channel at Boulogne. Said to have been built by Caligula, it was later restored by Charlemagne but collapsed into the sea between 1640 and 1644 (Photo 42).

Slow too is the gradual drowning of the land on which cities are built. We shall go further into this a little later as it is a phenomenon that has hardly been quantified and its medium-term consequences are not well understood.

In truth, human nature is such that it is much more impressed by sudden calamities.

The action of the weather, of the seasonal flow of rivers and of vegetation, or the creeping erosion of caverns and coasts are all imperceptible; man does not notice that his constructions are slowly but surely disappearing. Time passes and things

Photo 42. Engraving of the Tour d'Ordre at Boulogne-sur-Mer before its collapse into the sea around 1640. Top left: the tower as it was when besieged by the English in 1544.

change too slowly to be remarked during a mere lifetime. In contrast, how astonishing it is that certain geophysicists confidently offer maps showing the shape of our continents in 50 million years time! We learn that North and South America will by then be separate islands, that the Persian Gulf, like the Strait of Gibraltar, will have ceased to exist and that the Mediterranean will have been reduced to a smallish lake. The fact that we still do not know much about the forces driving the tectonic plates responsible for these changes makes their self-assurance all the more astounding.

But where will the coastlines be in fifty thousand years, or even in fifty centuries? Even though present trends suggest that certain cities by the sea are in real danger, no one is willing to stick his neck out. The more real the danger, it seems, the less people like to offer predictions: there is the glamour of a star-system for long-term prophets but a conspiracy of silence about what the immediate future, speaking in geological terms, holds for us.

After a look at the past history of submersion, which can help us to imagine some likely consequences in the near geological future and what man can do about them, we shall turn to those sudden disasters in the form of earthquakes, which shake the earth's crust and everything upon it, and try to glimpse the art of tomorrow's builders as they struggle to resist.

* 15 *

Submersion by the sea

'Vidi ego quod fuerat quondam solidissima tellus
Esse fretum; vidi factas ex aequore terras.'

'I saw the land formerly firm, turn into the sea and I saw land form where once was the sea' (Ovid (43 B.C.-18 A.D.) Metamorphoses).

And what Ovid had remarked had also been noticed by Aristotle: 'The same regions do not remain always sea or always land, but all change their conditions with time'.

Every coastline reflects a certain equilibrium between the level of the sea and the level of the land, and this equilibrium changes over time. If the level of the sea rises in relation to the land, the land is submerged and the coastline retreats.

Since the time man became sedentary some ten to twelve thousand years ago the level of the sea has risen about 60 metres, drowning a wide band of civilization. What do we know of the culture and technical skill of all those generations who built upon land that is now covered by the sea?

Figures 92 and 93, after Jelgersma (1979), show the shorelines of the English Channel and North Sea 10300 and 8700 years ago. The first human settlements were built on the banks of the Rhine and the Thames, two rivers which joined forces before turning south to reach the sea north of Normandy.

A number of authors have attempted to chronicle this gradual rise of the sea. One thing on which they all agree is that the rate of submersion has slowed down in the last few millennia: from two metres per century some 7000 years ago it has now fallen to one or two decimetres (Figure 94).

In mesolithic times, the rising level of the sea endangered only the temporary settlements of the ancient hunters and gatherers, some of whom were just beginning to till the land. Today, however, the continuation of this trend – even at a slower rate – would threaten the very existence of many great cities. The fact that Unesco has recently launched an investigation into sea levels shows that people are aware of the danger. Only the low-lying coastal areas are affected, especially where the phenomenon of submersion is aggravated by tectonic movements. As we shall see, the rate of submersion varies with the sea concerned.

We shall take a few examples of man's struggle, both now and in the past, to protect his constructions against this invasion by the sea.

THE WESTERN MEDITERRANEAN

It was long thought that the coastlines of the Mediterranean had never moved (Cayeux 1907) but a group of archaeologists, geologists and geographers have recently published a study demolishing this theory. The evidence they produce is strong since certain telltale discoveries of underwater archaeology, such as the base level of ancient stone quarries, the entrances to tombs located on the sea-shore, the stone quays of some early ports, or the sluice-gates of the fishponds which flourished in Roman times, are quite precise.

For instance, the Roquetaille Quarry at Cape-Couronne near Martigues (France) was, as Strabo relates (IV.1.6), exploited by the Romans, who cut into the rock to extract the limestone formed during the Burdigalian stage of the Miocene. The bottom of this quarry is now under 30 cm of water

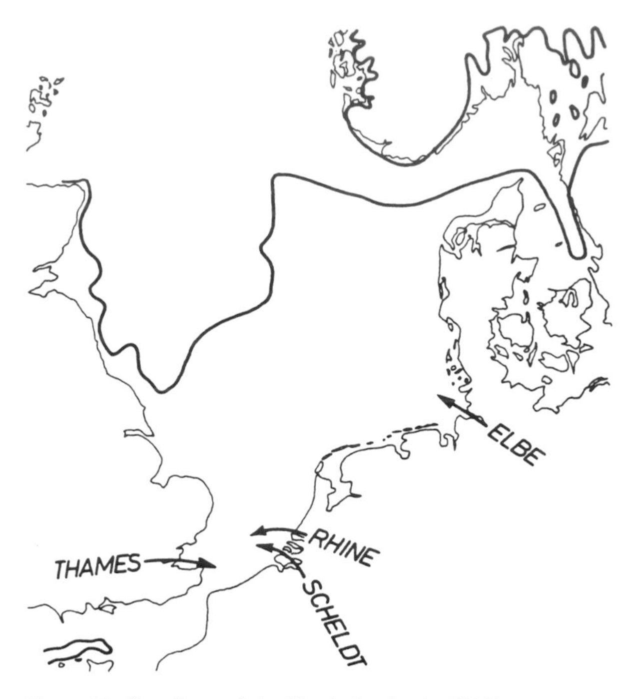

Figure 92. Shorelines of the North Sea basin 10300 years ago, when the sea level was about 65 m lower than it is today. After Jelgersma 1979. The British Isles were still attached to the continent. The thinner lines represent the present-day coast.

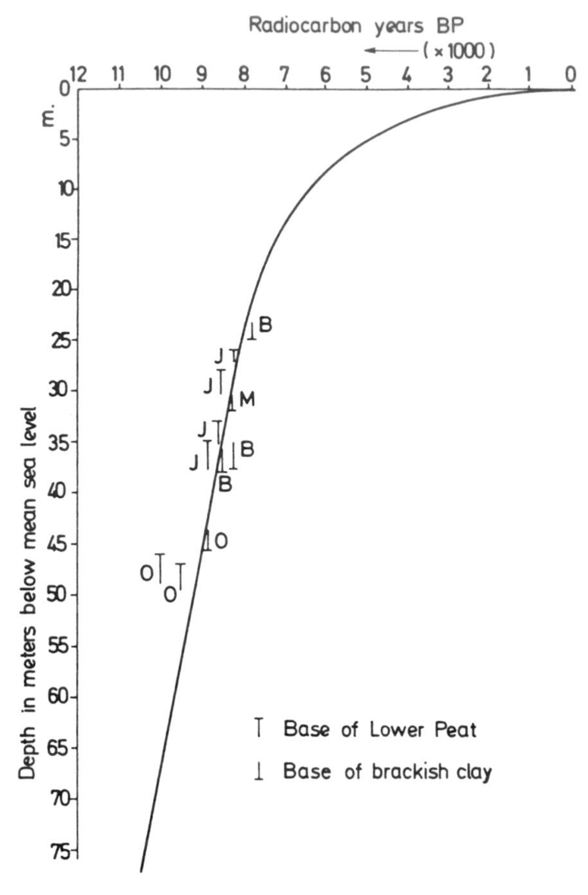

Figure 94. Curve for the relative rise of sea level during the last few thousand years (mainly based on data from the Dutch sector). After Behre, Jelgersma, Morzadec, Oele. 1979.

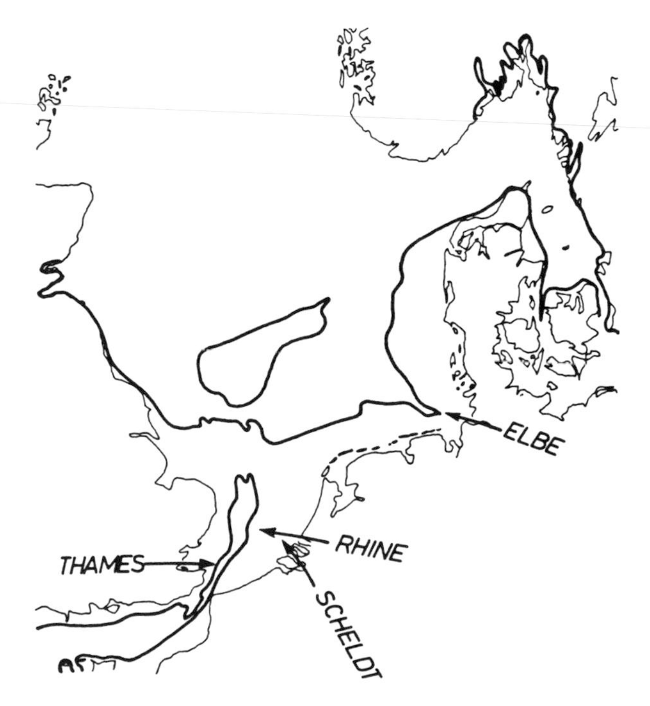

Figure 93. Shorelines of the North Sea basin 8700 years ago, when the sea level was 36 m lower than the present level. After Jelgersma 1979. The thinner lines represent the present day coast.

(Guéry et al. 1981). The same thing has happened to another quarry at R'Mel, east of Bizerta (Paskoff et al. 1981), which used to provide Carthage with sandstone.

At Mahdia Gummi (Tunisia) there exist some eleventh century Muslim tombs which are now in the tidal zone (Paskoff et al. 1981). Some other tombs, hewn out of the rock of abandoned quarries at Ra's Amir, the northernmost headland of Cyrenaica (Laroude 1981), are now completely submerged.

Ports

In 1967, excavation work prior to building an underground carpark for a new shopping centre near the Bourse at Marseilles confirmed that the sea level has risen somewhat since Roman times when it uncovered the ancient Roman port of Massilia just east of the Greek ramparts (Figure 95). This port extended northwards along a valley bottom which now carries the 'Canebière', the main north-south artery of the present-day city. The substratum is an eroded Stampian marl overlaid with silt and gravel saturated with water owing to the proximity of a very abundant spring, known as the Lacydon, which was only about 30 m from the northern end of the port.

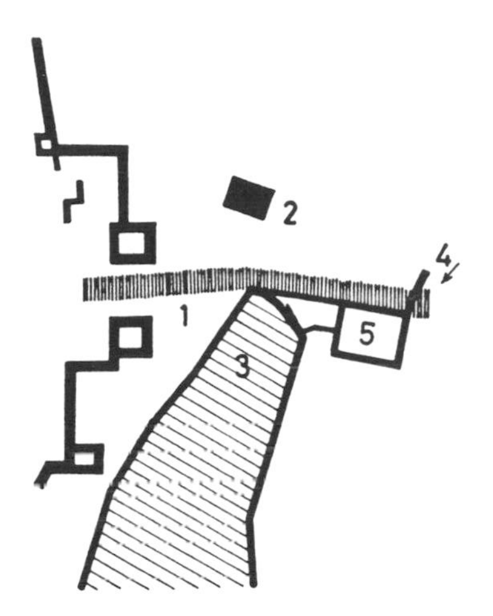

Figure 95. Plan of ancient Massilia, based on the excavations in the area of the 'Bourse'.
1. Via Italia 2. Ancient Greek ramparts 3. Harbour 4. Aqueduct from the Lacydon spring 5. Basin

The Romans built the walls of the quay in the dry by using the method described by Vitruvius, i.e. behind a barrier of compacted clay held in place by two parallel walls of slotted planks (see Photo 36). The pile planking recently discovered on the site for the underground carpark was formed of squared wooden posts, pointed at the top and with vertical grooves to hold the horizontal planks. The posts were driven more than 3.50 m below sea level, into the Stampian marl. Long wooden planks were then slotted into the grooves and the water on the inside pumped out, so that the Romans could build their port in dry conditions.

The quality of the work was praised by the Latin poet Avienus in about 695-699 A.D.: 'Thus has the hand of man compelled the deep sea to penetrate into the land and the perserverance of the ancient founders of the city won a skilful triumph over the form of the site and the nature of the soil'.

The original walls of the quay, with not a stone missing, were formed of six courses of stonework 50 cm high, making a total height of 3 m. The five upper courses contain stones of from 1.20 m to 1.70 m long, which were laid on a line of bigger stones placed crosswise and toed out from the wall to give greater stability. The face of the wall is still perfectly vertical (Figure 96). At a later period, during the Lower Empire, three further courses were added to the original six.

In the valley-bottom north of the apex of the

Photo 43. The carefully dressed quay wall of the ancient port of Massilia. The top three courses of stonework were added much later, during the Lower Empire. The black stain running along the fourth course from the top shows the ancient sea level (Photo CNRS – A. Chéné and G. Reveillac).

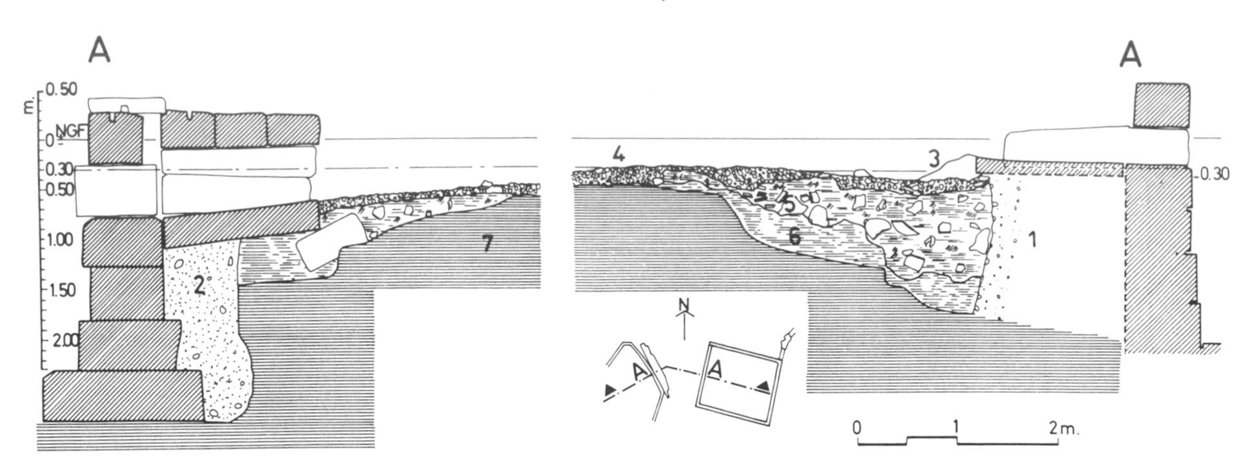

Figure 96. Schematic cross-section between the basin and the harbour.
1. Concrete 2. Foundation trench 3. Sand 4. Layer of pebbles 5. Broken amphora 6. Silt 7. Stampian substratum.
(Drawn by J. Lenne from survey by Borély and Guéry, CNRS).

Photo 44. The layers of broken amphora (Photo G. Bertucchi, National Archives).

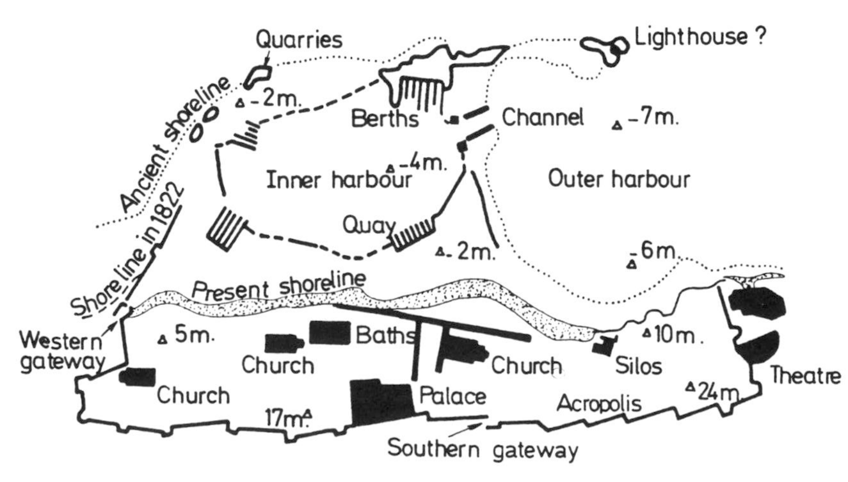

Figure 97. The ancient port of Apollonia (after Laronde 1981).

Port (Figure 95) lay a marshy area which had been filled in to make way for the Via Italia, on the northern side of which were burial grounds. The section shown in Figure 96 reveals the subtle means used by the Romans to strengthen the silt beneath the filling: they placed a layer of broken amphoras (Photo 44) between the silt and the fill to absorb the pressure of the water squeezed out of the silt by the weight of the fill.

The ancient sea level, indicated by the dark stain (Photo 43), is just over 40 cm below the present sea level.

In Cyrenaica, on the opposite shore of the Mediterranean, the ancient port of Apollonia (today called Susa), lay next to the Roman town (Figure 97). It had an outer harbour giving access, via a narrow channel flanked by two towers, to the quays and berths of the inner harbour. In front of the central and western parts of the Roman town, the shoreline appears to have retreated southwards by about 110 m and the ancient port is now under 2-3 m of water. The sea has also invaded some ancient silos cut into the rock at the foot of the Acropolis.

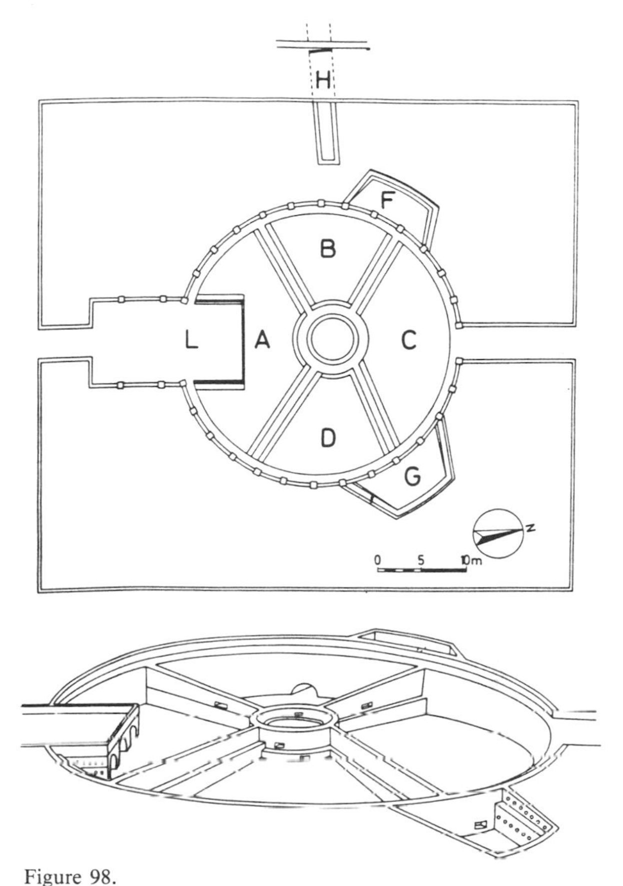

Figure 98.
Above: The fishpond of Lucullus. Diagram.
Below: Oblique view from the east, after Chiapella.

Fishponds

The Roman fishponds date back to the first century B.C.. At the time they were built, the tops of their surrounding walls were above sea level, so their present depth under water gives a rough indication of how much the sea has risen. Moreover, the saltwater in the fishponds was renewed by means of two sluices cut into the rock just deep enough to allow the water to enter at high tide. The operation was controlled by a system of sluice-gates (called 'cataractae') which were raised or lowered in grooves. The present depth under water of the bottom of these channels thus gives another clue to the sea level in ancient times.

During a survey carried out for the Italian Research Council, Schmiedt (1981) examined 8 of the 22 fishponds along the Tyrrhenian coast.

The fishpond of Lucullus (Figure 98), now entirely submerged, had a most ingenious design. It was circular in shape, with a diameter of 32.4 m, and divided into four unequal compartments (A, B, C, D). 'H' indicates the inlet channel. A platform built on vaults (L) projected into one of the basins, providing a cool shaded area for the fish to shelter from the summer heat. The pools marked 'F' and 'G' were employed for raising certain types of fish, which bred in decapitated amphora. Clearly, the Romans put old amphora to a surprising variety of uses!

For each of the 8 fishponds he studied, Schmiedt has given figures showing that the sea level has risen by an average of 60 cm, with a minimum of 55 cm and a maximum of 105 cm. Other observations from underwater archaeology in the Western Mediterranean also converge on this average figure of 60 cm, with differences probably due to subsidence or uplift here and there. It is a moderate though genuine rise in the course of twenty centuries. It is by no means certain, however, that the pace has been uniform (i.e. 3 cm per century). Indeed, there is some evidence to suggest that the rate of submersion is now greater: at the Anse Calvo in Marseilles, for instance, the average level of the Mediterranean measured over the ten years between 1968 and 1977 is 11 cm higher than when measured over the 12 years 1885 to 1896. Unless the soil at Marseilles has settled an exceptional amount between these two periods, it would seem that the rate of submersion has now tripled in the Western Mediterranean. Present-day estimates suggest that the water in the northern hemisphere is now rising by about a decimetre per century; where the rate of submersion differs from this, the explanation is that tectonic movements are affecting the level of the land.

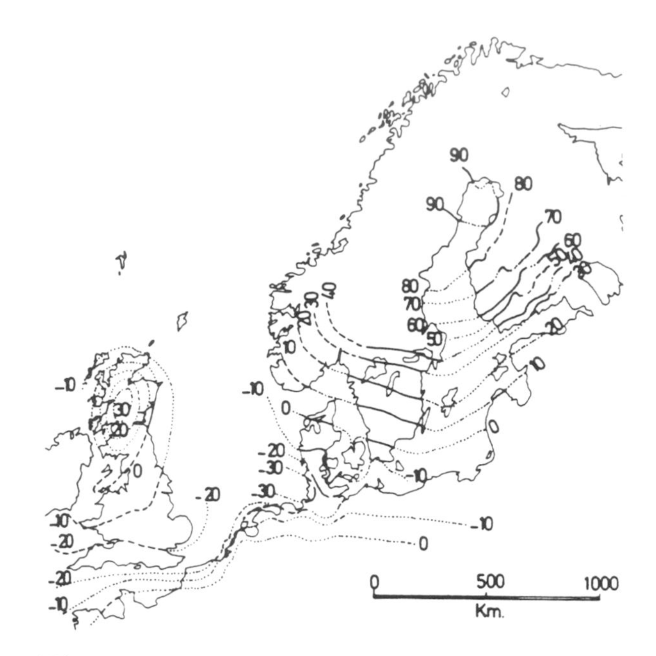

Figure 99. Present uplift and downwarping in northwest Europe. The isobases represent the rate of change in cm/century, determined from tide-gauge records. Dashed isobases are less certain and the dotted isobases are based on interpolation (from West 1968).

FROM THE GULF OF BOTHNIA TO THE GULF OF GASCOGNE

The most recent glaciation lasted 80000 years and ended 20000 years ago. It froze the rivers, preventing their water from reaching the sea, and imprisoned vast quantities of water in immensely thick icecaps. When these icecaps melted during the subsequent deglaciation, which is still going on, the sea rose and the mountain-bearing land was relieved of the weight of the ice.

As we have seen, the Sumerians, the Aztecs, and the Venitians took advantage of the loads they placed upon their soft soils: when overloaded, these soils became stable; what is more, they did not tend to rise once the weight had been removed. It was as if they remembered the heavier loads they had previously had to bear.

With the mountains, however, it was quite a different matter. When relieved of the enormous weight of the ice, they started to rise because they float, so to speak, on a viscous liquid with their roots as deep as 50 to 60 km in the earth's crust. With the melting of the ice, the whole system rose together, like a ship when its cargo is unloaded.

An example is provided by the northern shores of Hudson Bay, where the land has risen by at least 9 m – about 1 m per century – since the Thule Eskimos settled in the region. As one can see from Figure 99, certain beaches at the northern end of the Gulf of Bothnia are today uplifting at about the same rate. The map also shows the more modest uplift in Scotland as a result of the melting of the icecap which once stretched from Iceland down to the latitude of London.

During the last glaciation, the mountains were weighed down while the zones on the periphery of the glaciers were uplifted. Today, however, this seesaw movement is acting in reverse: the peripheral zones are now subsiding (Figure 100) and the effect is amplified by the denudation of the mountains formerly covered by the icecaps and the deposition of glacial sediments on the rim.

This subsidence, or 'downwarping', is shown on West's map (Figure 99). We can see, for example, that certain parts of the coast between Denmark and Holland are subsiding by 30 cm per century and that southern England is subsiding at around 20 cm per century. West appears to think that the Channel Islands, the Cotentin and North Brittany are not affected but his deductions from the evidence seem rather conservative: for example the Roman historian Strabo relates that it was possible to walk from Cotentin to Jersey! In Guernsey, and in Cornwall too, the remains of ancient forests alive less than two thousand years ago can still be perceived at low water or just beneath the low water mark. According to a manuscript conserved in the Library of Avranches, written by one of the first Canons of Mont St. Michel, the forest of Scissay, in the Bay of Cotentin just north of Mont St. Michel, was not submerged until the eighth century though analysis by Carbon 14 does in fact suggest an earlier date. Lastly, to the west of Mont St.

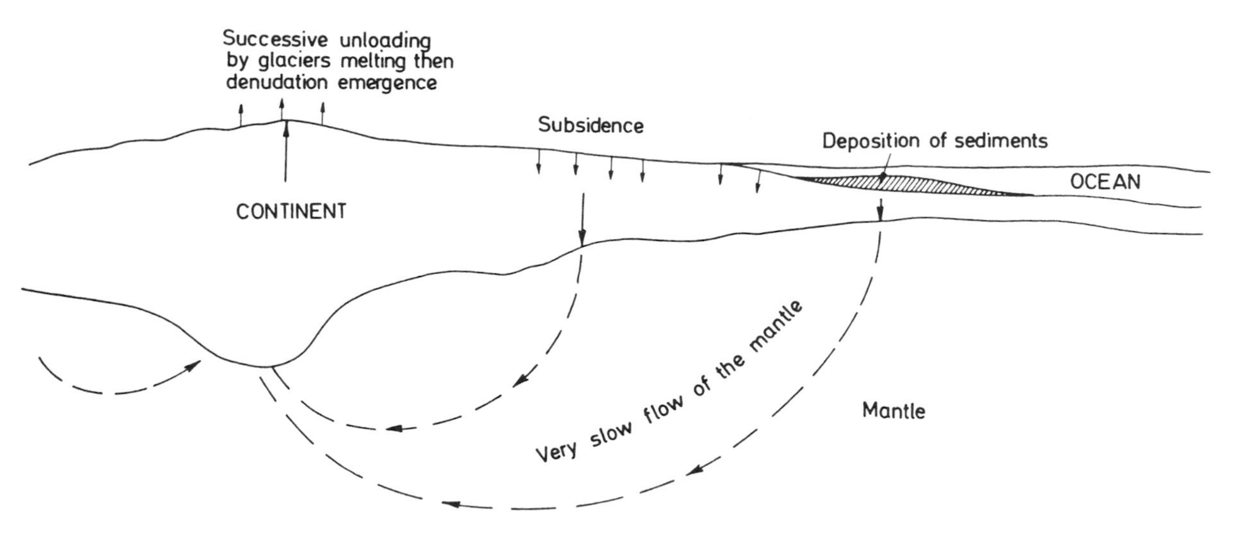

Figure 100. Uplift and downwarping during a period of deglaciation (accentuated by denudation of reliefs and related deposition of underwater sediments). The vertical scale has been greatly exaggerated.

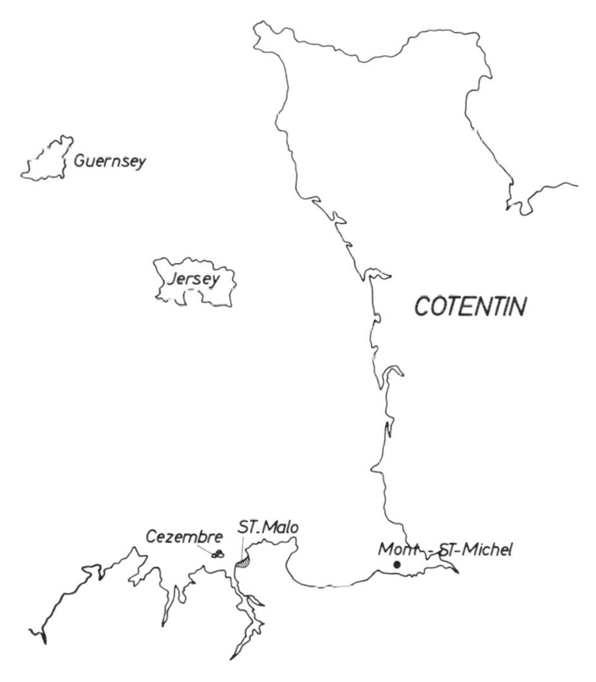

Figure 101. Map of the Bay of Cotentin.

Michel, in the 'Rade de St. Malo', (Figure 101) the submarine plateau between St. Malo and the Island of Cézembre was still dry land partly used as pasture in the sixteenth century: the archives of the Ile-et-Vilaine Department contain references to transactions and disputes concerning 'the pasture and island of Cézembre' in 1486, 1506, 1507, 1510, 1511, 1515, and 1516. Clearly, the submersion of the last two thousand years or so is still going on and has been far from negligible in the last four centuries. Indeed, as in all zones on the rim of ancient glaciers, the land is subsiding while the sea level is rising.

The battle against submersion

How did the people living on low-lying coasts fight back?

In the North Sea, few remains of early prehistoric settlements have been found because the sites have been buried by sediments. Those which have been discovered on the bed of the North Sea reveal that in mesolithic (12000 – 7 000 BP) and neolithic (7000 – 4 500 BP) times there existed colonies that grew cereals. They lived on 'wurts' ('wurten' in German and 'terpen' in Dutch), which were natural mounds that were gradually built up higher and higher to keep out of the range of storms.

There were, however, moments when the submersion paused for a while. Around the time of Christ, for example, the sea even retreated for a short period (what geologists call the Dunkirk period), and villages began to extend beyond the coastline on to the marshy areas. In the first century A.D., the sea level began to rise again and these settlers were forced to build up the soil in stages in order to construct new dwellings on top of the older ones. Figure 102 represents the cross-section of a 'wurt' dating from this period, showing the various layers of

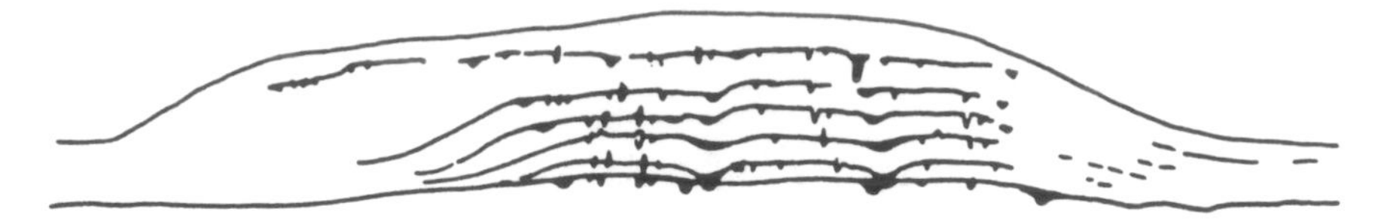

Figure 102. Schematic representation of a wurt of Roman time. At the bottom a natural levee with a flachsiedlung followed by several clay layers deposited by man, each with a habitation layer containing the foundations of houses etc. (from Behre et al., 1979).

habitation built up to form an artificial mound that eventually reached a height of about 7 m above sea level.

The first level of buildings was on the marshy soil ('flachsiedlungephase'). In excavating the Wurt of Feddersen Wierde, Haarnagel discovered seven complete villages one on top of the other. Their levels, adjusted to allow for settlement, give an idea of the sea level during storms, while analysis of the pollen in the soil tells us what crops were grown and approximately when.

The 'wurt' period ended towards 1000 A.D. with the development of earthworks and dykes to carry dwellings and protect the farmland behind. When this was done in estuaries, however, the sea had insufficient room to extend, rose and sometimes broke through the dykes. Later, attempts were made to increase the fertility of the protected land by drainage – but this lowered the level of the land, causing the dykes to fail or making it necessary to strengthen them.

Threatened coastal cities around the North Sea

In the past, man threw up earth works to protect himself from the sea. What sort of action is he taking today?

United Kingdom

The subsidence of southeast England together with the slow rise of the sea level have made the London area a potential scene of disastrous flooding should the highest tides happen to occur at the same time as a strong surge.

Figure 103 plots the 8 exceptionally high tides recorded at London Bridge over the last two hundred years (Gilbert and Horner 1984). The flood in 1953 covered some 80000 hectares and caused the death of over 300 people. In Holland, the same exceptional tide left over 3000 dead.

In Figure 103, the line formed by plotting the

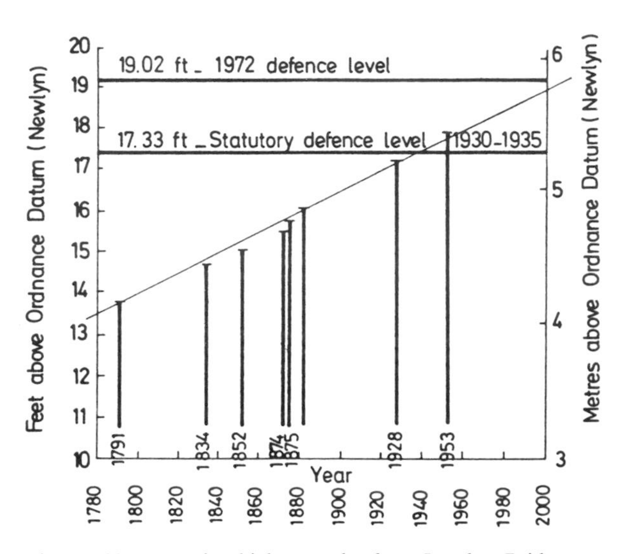

Figure 103. Increasing high-water levels at London Bridge.

peak levels of these tides suggests that the maximum is increasing by about 75 cm per century. Meanwhile, the land is sinking, not only because of the already mentioned melting of the glaciers but also because of the loads that the thick layer of clay in the London region has to bear. If, moreover, the highest tides coincide with an atmospheric depression, the resulting surge in a V-shaped estuary can raise the level still further.

When it was realized just how disastrous for the London area would be the damage caused by a tide only a little higher than the one in 1953, it was decided to construct a 'barrier' on the bed of the Thames at Woolwich Road, about 6 miles east of Westminster (Figure 104). When everything was normal it would remain in its housing but could be raised into position at the approach of a storm. The Thames Barrier, completed in 1982, has four gates 61 m wide and six others 31.5 m wide (Figure 105); two of the latter pivot like the big ones (rising sector gates) while the other four are raised or lowered. The project as a whole has cost some £440 million.

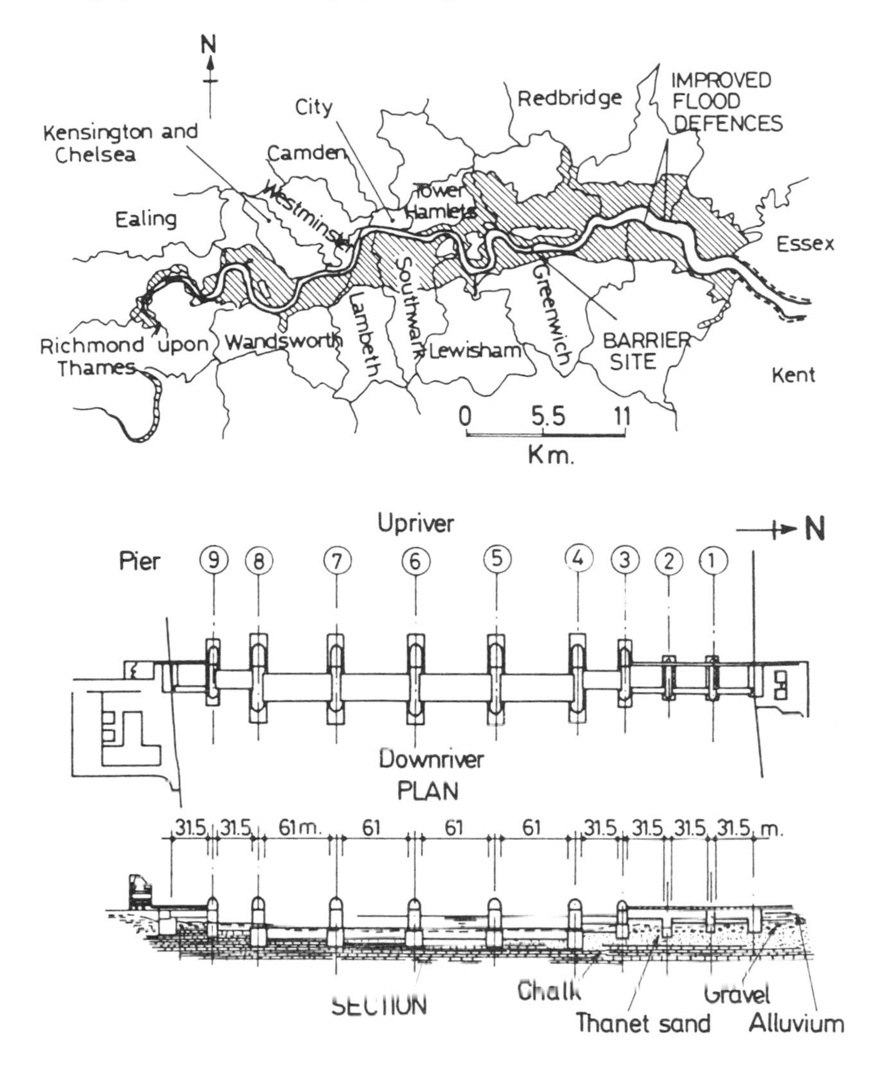

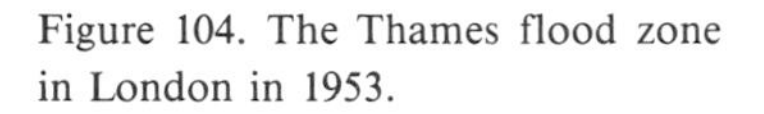

Figure 104. The Thames flood zone in London in 1953.

Figure 105. Cross-section of the Thames Barrier.

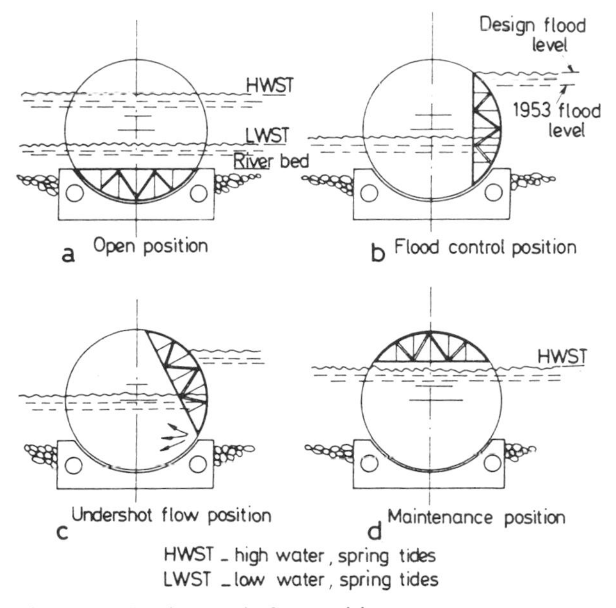

Figure 106. Barrier gate in four positions.

Netherlands

The history of the Netherlands is that of the Rijkswaterstaat's long and successful struggle against the encroachment of the sea and the constant need to choose between maintaining the Rhine estuary and protecting the land, since it was and is impossible to do both. In this case, the sinking of the land is due to two factors: the periglacial location and the settlement of the alluvium deposited by two powerful rivers.

As part of the so-called 'Delta Project', which has just been completed, it was decided to close off the estuaries and channels in the southwest of the country except for the ones giving access to the ports of Rotterdam and Antwerp. The plan involved six earth dams and two barriers which could be shut in the event of storm surges. The larger of these, the Oosterschelde Barrier, involves 65 enormous piers, each weighing some 18000 tons and ranging in height from 30.25 to 38.75 m, set 45 m apart, with sliding steel gates that can be raised and lowered between them.

When the barrier is closed, these piers have to withstand great lateral pressures and it proved very difficult to design adequate foundations on a compressible and erodable soil. The Dutch engineers (Netherlands Commemorative Volume 1985) have overcome the problems by using some of the most advanced and effective techniques in soil mechanics. For instance, they have compacted the riverbed to a depth of 15 m by means of giant vibrating needles and, on this improved surface, spread foundation 'mattresses' 200 m long by 42 m wide and 0.36 m thick under each pier. On each of these bottom mattresses a second smaller mattress (60×29×0.36 m) was then laid. These mattresses are in fact giant cushions made of artificial fibres and filled with sand and gravel. They are therefore permeable and able to absorb the changing water pressure in the subsoil induced by the tidal action.

Germany

In 1962, the tide at Hamburg rose 4.20 m higher than expected. The consequent flooding affected 120000 people, of whom 312 died. Since then, the banks of the river have been raised and a raised walkway constructed in the centre of the city, but the problems of the port and naval dockyards still remain to be solved. Here too they are thinking of a barrier operating on the same principle as the one in London.

U.S.S.R.

Leningrad is built at the head of a long and narrow sea and criss-crossed by canals. From time to time, when the wind blows from the Baltic, the sea level rises and causes flooding in the streets. This has happened 250 times since 1703. The Neva, a large river with immense docks and quays along its banks, flows through the city. In this case, the plan is to build, beyond its mouth, a vast dyke 25 km long with locks for the ships. Its construction is now well under way and it is expected to be fully operational in 1990.

Already then, the great ports of the North Sea are under siege by the sea and man is busy preparing fortifications like invisible underwater drawbridges that can be raised when the attack comes.

But they are fortifications with a limited life: Gilbert and Horner (1984) think that the Thames Barrier will have to be raised at least twenty times a year in the next century and that the gates will last less than two hundred years. The settlement of the London clays will, of course, diminish with time, as might the other factors. One can only hope that this will be so for if the present trend continues – if, for instance, we prolong the line in Figure 103 to the year 14000, i.e. twice the time that separates us from the end of the Palaeolithic, it shows that London would be under 120×0.75 m, that is 90 m, of water.

Each interglacial period has been marked by shorter periods of a colder or warmer climate which slow down or speed up submersion. Even though the situation in 12000 years time will depend essentially on the shift in the climate between now and then, it is not at all easy to say whether we are now on the verge or at the end of one of these sub-periods.

Mohenjo-daro: A hopeless struggle against submersion

During the nineteenth century, British, American and Indian archaeologists working in the Indus Valley, in what is now Pakistan, discovered over 150 sites of ancient towns stretching right up to the foothills of the Himalayas. One of them, nearly 300 km from the coast, is Mohenjo-daro.

The civilization which once inhabited this city has raised many problems for archaeologists. A pottery bas-relief discovered in the ruins portrays what appears to be a sea-going boat. Elsewhere, in the ruins of certain Sumerian towns, clay seals originating from Mohenjo-daro have been found, suggesting that ships from the Indus used to sail as far as the Persian Gulf. It was also long thought that the superimposed cities discovered on the Mohenjo-daro site had been submerged by exceptional floods of the Indus, which buried the ruins of one after the other under vast quantities of clay and silt. Indeed, the deposits covering the earliest buildings are at least two metres thick. The theory was that the inhabitants obstinately reconstructed their city on the mud left behind after each flood, and it was remarked that each time this happened the standard of building fell. The lowest levels reveal spacious buildings with solid walls which contrast with the makeshift dwellings of the upper ones – as if the inhabitants gradually lost heart.

R. Shani, an Indian scholar, R. Raikes, a hydrologist, and R.H. Dyson, an archaeologist, have recently proved that the real reason for the decline of Mohenjo-daro was not submersion by the overflowing Indus.

Though it is true that the lower levels of Mohenjo-daro are embedded in some 10 m of silt deposited during the 2000 years of the city's existence, it is odd that there is no further trace of silt for the subsequent 3500 years. Moreover, a detailed analysis of the layers of silt has revealed that they are too thick and spread too evenly to have been deposited by the flooding Indus. On the contrary, they would appear to have been deposited by an enormous body of stagnant muddy water. At the same time, it was noticed that the towns discovered in the Indus Valley had a peculiar feature in common: their ramparts seemed to be directed against invasion from the direction of the Sea of Oman.

Raikes, impressed by the seaward-looking defences of these Indus Valley settlements, concluded that they used to be coastal ports. Why were they now so far inland? Careful study of aerial photographs revealed that there existed not a single archaeological site between the present mouth of the Indus and a point some 140 km inland; all the sites were further upstream. The photographs also pinpointed at least 10 prehistoric beaches, parallel to the present coastline, that could be recognized from the concentric curves visible from the air. Some of these ancient beaches were well inland to the north.

It was eventually concluded that a tectonic movement had uplifted the whole of the coastal plain through which the Indus now flows on its last 140 km before reaching the sea. This movement was accompanied by eruptions of underground magma and mud – quite a common phenomenon in that region (a mud volcano erupted at Hyderabad in 1819, creating a fold that obstructed one of the tributaries of the Indus and drowned some 5000 km^2 of land under stagnant water for two years, while another mud volcano in the Sea of Oman in 1945 caused islands to emerge). In the case of Mohenjo-daro, a wave of similar eruptions had partially blocked the course of the Indus, forming a huge lake which stretched up to Mohenjo-daro and beyond. Thus the city suddenly found itself isolated in the middle of a slowly rising expanse of swampy water.

When the 140 km wide band of continental shelf suddenly emerged from the sea, Mohenjo-daro became twice as distant from the coast as before. This also explains why there are no archaeological sites along the lower reaches of the Indus. What a perfect example to illustrate the remark of Aristotle quoted at the beginning of this chapter!

The remains of Mohenjo-daro bear witness to the gradual development over several thousand years since the founding of the first cities such as Jericho of what we now call town planning. Though the layout was rather monotonous, with its long parallel lines of houses, it offers the first break with an architecture based on a religious or civil hierarchy. The attempt at planning led to advances in a highly efficient system of underground drainage and more generally in all the applied arts.

Yet this entire civilization was wiped out, not by the decadence which has destroyed so many brilliant cultures but because the people lost heart and finally gave up the hopeless battle against submersion. Eridu, after a long struggle, eventually achieved a firm foundation but Mohenjo-daro was doomed to be engulfed.

Delta cities in danger

Venice

One of the best films by Visconti is called 'Death in Venice'. Today, however, it is the death of Venice, the most powerful city of medieval Europe, that threatens. As we have seen, this city of art was built by a series of remarkable architects on a soil that was almost like quick-sand.

Its 'death' has long been predicted by travellers. In 1929, André Suarès called it 'a preposterous city without land, in which a hundred thousand piles carry a church with the sleight of hand of a juggler'. In 1864 Hippolyte Taine spoke of the host of shabby deserted streets 'in which the pile foundations are encrusted with shells and become so thin that the buildings could collapse at any time'. For Dominique Fernandez (1980) 'The sea at Venice is the colour of tar and pitch... a dark gullet ready to swallow and bury in its silt the foolhardy constructions of human kind'. These conditions are already killing the city by desertion: in 1950 the historic centre had 185000 inhabitants but less than 100000 25 years later.

At the beginning of the century the Venetians, convinced that their city could no longer continue in the shadow of its former glories, decided to build

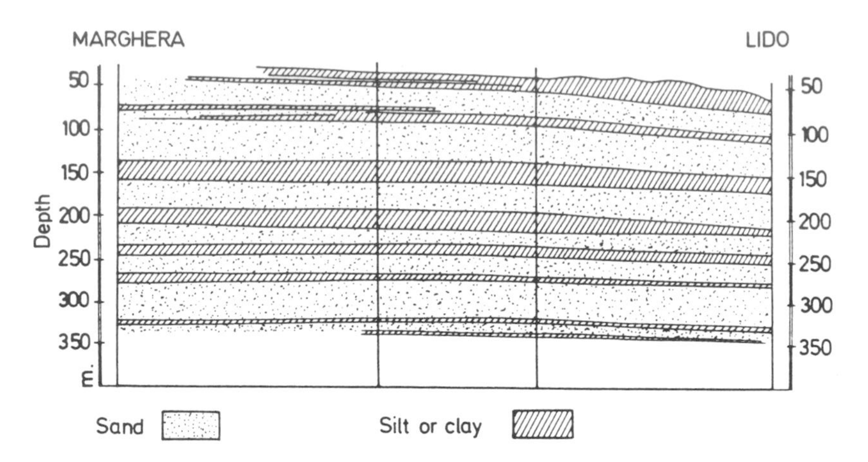

Figure 107. Schematic representation of the lagoon subsoil (depths given in metres).

Mestre and the port of Marghera on the mainland. The two cities quickly expanded, creating water needs which were met by pumping from the layers of sand interbedded in the silt deposited by the Po, the Adige and the Piave at the time the lagoon was formed. This uncontrolled pumping has provoked the settlement of the silt with the result that Venice, the victim of its desire for a new lease of life, now suffers from what Venetians call the 'acqua alta', i.e. regular flooding by the water of the lagoon. The pumping continued for over fifty years, inducing a settlement of 13 cm (Ghetti and Batisse 1983), before it was stopped in 1975. But that will not save Venice.

In the first place, the damage has been done and the lower parts of the buildings are being attacked by polluted water. Secondly, the many layers of silt (Figure 107) will continue to settle under their own weight, slowly of course, but still at about 0.5 to 1 mm per year. Thirdly, the water level is rising by about one millimetre per year. In addition, the most recent ice age, the Wurmian, descended a long way from the Alps so that after the melting of the glaciers the land, as in any peripherical zone of an unloaded region, is tending to sink. In sum, the records of the tidal gauge at the Salute headland at Venice show that the sea level rose by 26 cm between 1870 and 1980. But these records concern only the mean level of the tide; the spring tides can attain an amplitude of 90 cm, plus up to 20 cm more owing to low pressure zones during the winter. Lastly, as Venice lies at the head of the long and funnel-shaped Adriatic, the Sirocco, which blows from the south, is capable of increasing the effects of the tide. If all these factors combined their maximum forces, there would be 2.50 m of water on the

Photo 45. 4 November 1966: the Piazza San Marco under 1.96 m of water. (Photo by Cameraphoto).

Piazza San Marco. Indeed, on 4 November 1966, the water was 1.94 m deep even though it was not a spring tide. Since then the Piazza San Marco has been badly flooded twice, to a depth of 1.66 m in 1979 and 1.59 m in 1986.

In 1975 an international competition was launched to find a solution, but none of the five proposals was accepted.

The project officially adopted in 1976 involved

Photo 46. Water pouring over the 'murazzi' protecting the islands of the lagoon. The photo was taken on 4 November 1966 at 9:30 a.m. It is waves such as these that the barriers to protect Venice would have to withstand. Photo by Ermanno Reberschak.

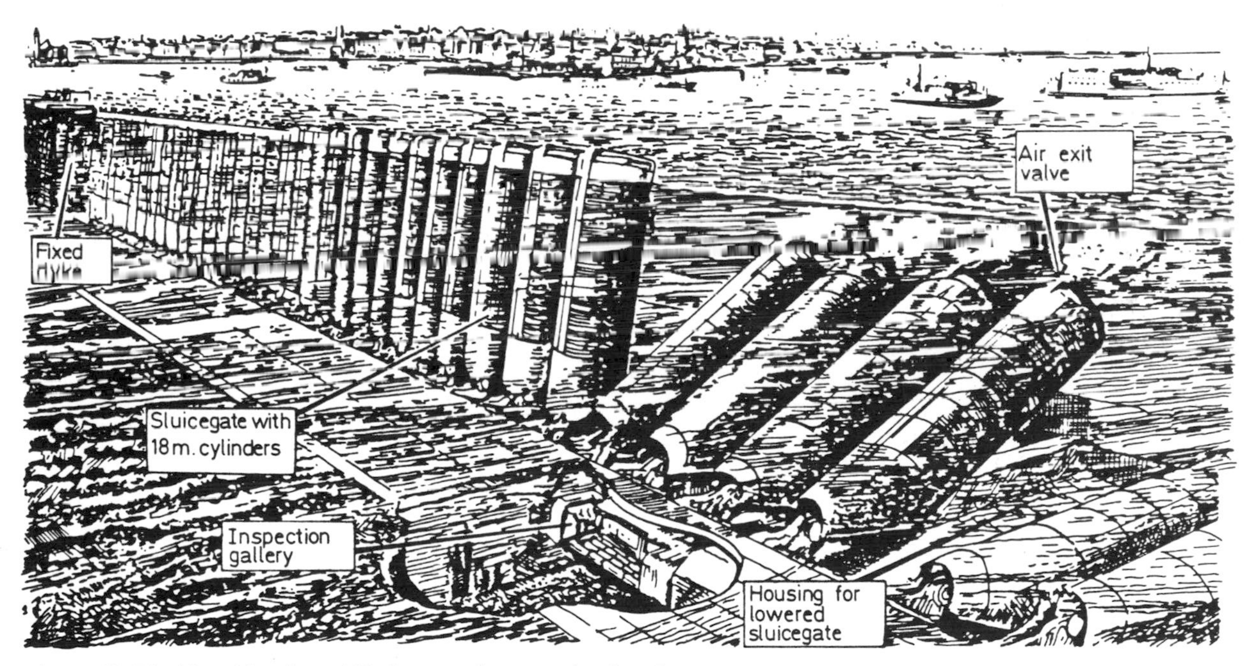

Figure 108. The hinged barriers of Venice, a project now abandoned.

blocking off each of the five channels between the lagoon and the sea with barriers that would be raised from their housings on the channel-bed in case of danger from storm and/or tide. But some of these channels are 300 m wide and the barriers would have to be 18 m high to cope with a storm surge – a much more difficult problem than in the case of London or the Rhine delta.

What the Venezia Nuova Committee had in mind was a long line of hollow cylinders (Figure 108) lying on the seabed and hinged at one end to a horizontal axis (which would have to be very firmly anchored); they were to be raised by having air pumped into them. But the project has now lost favour for several reasons.

The main problem is that some of the 'acque alte' are minor and some very big. The minor ones are very common – the critical level of 70 cm (at which water begins to appear in Piazza San Marco) was, according to Ghetti and Batisse, exceeded over a thousand times between 1970 and 1979. If the barriers were raised on each occasion they would not only be a great nuisance to shipping but would disrupt the natural flow of clean water that flushes away all the filth poured into the lagoon.

The city of the Doges is doomed: it will slowly expire, deserted by its inhabitants, following the same fate as Torcello, the first island in the lagoon to be inhabited, which had a population of 20000 at the beginning of the century and whose splendid stones, like the Tura limestone used to face the pyramids of Giza, have been removed for use elsewhere.

All our great-grandchildren will probably see will be a museum city bereft of its marbles and sculptures, removed by the retreating Venetians to adorn their new residences on firmer soil.

Bangkok

The site of Bangkok has a lot in common with Venice. The capital of Thailand is built on the largest rice-growing plain in South-East Asia, whose fertility accounts for its population of five million inhabitants. As it grew, however, the city quickly spilled on to an area of highly compressible soft clay which is now sinking under the weight of the buildings, railways and roads. It has almost reached sea level and, as in many other places, water is being pumped from the sandy aquifers beneath the clay.

Rau and Natalays (1983) have recently calculated that if settlement continues at the present rate Bangkok will be at or below sea-level in 10 to 20 years. Fertility and urbanization are proving incompatible without discipline: if nothing is done, the city will become a lake with its buildings on piles emerging from a steadily sinking soil.

* 16 *

Sudden destruction: Earthquakes

Earthquakes and landslides have their place among the forces that have erased the traces of a civilization. From the remotest times they have caught the interest of numerous writers. The engulfing of Helice, to take but one example, is recounted by Heraclides Ponticus, Strabo, Pausanias, Aelian, Ephorus, Diodorus, and Latin writers such as Ovid, Pliny and Seneca.

Helice was an Achaean town that must have been very ancient since it is celebrated by Homer. It extended up to two kilometres from the coast on a site lying between two small rivers, the Selinos and the Kerkynites, which flow into the Gulf of Corinth (Figure 109). It had a powerful fortress built of cyclopean blocks and, nearby, a sacred grove sheltered the statue of a god, lord of the soil and the shore, whom the Greeks would later call Poseidon.

In 373 B.C., Pausanias relates (VII, 24.7), the sun shone with a reddish glow and flashes of lightning streaked across the sky. For five days, mice, snakes, rabbits and all animals that live in holes in the ground emerged from their burrows and took flight before the astonished gaze of the people of Helice (Aelian, On the Characteristics of Animals, Book XI, 19). Then one night, with a great roar, the town slid into the sea and disappeared; an immense wave swept over the shore and ten Spartan warships anchored off the coast were dashed to pieces against each other.

Earthquake or landslide? Nearly 2400 years later it is difficult to say for certain especially as even a light tremor can often trigger a landslide. In this case it would seem that both were involved since the rocky substratum underlying the site of Helice collapsed into the sea. In his analysis of slides in landforms of this type, Muller (1963) shows that the process begins with overloading at the foot of the cliff (Figure 110) combined with the gradual formation of cracks that spread upwards through the rock and eventually appear as fissures at the surface. The first tremors, unnoticed by the inhabitants, hastened the process and the noise of the splitting rock deep within the earth drove the terror-stricken animals from their burrows. (Today the Chinese have a theory that animals can simply detect the smell of an inert gas – radon – which is forced to the surface by the compression of the rock shortly before an earthquake). The zigzags of lightning heralded violent storms, the rain from which poured into the fissures, increased the pressure and aggravated the situation. Later, as we are told by Eratosthenes, travellers returning from that coast spoke of a bronze statue standing upright beneath the waves, a human figure holding a seahorse in his hand, which fishermen were afraid of entangling in their nets... The reappearance of the sacred grove and its statue of Poseidon accords perfectly with a landslide of the type described by Collin (Figure 57) and is very similar to what happened at San Fernando (Figure 71) or at Panama (Photo 28).

Helice had been a great city and the mother of colonies. In the Tarentine Gulf in southern Italy, a part of Magna Graecia at the time, lay Sybaris (Figure 111), a Greek city of over 100000 inhabitants, which had been founded by Helice around 720 B.C.. This city, too, has vanished, though remains of it have been found under the alluviums of the rivers that flow down from the Appenines southward through Lucania. Towards 1960, the Italians decided to build an industrial port beyond the site of these ruins; in the next few years two unsuccessful attempts were made to carry out this project but each time the new mole was swallowed up by the waves or, more exactly, sank into a sea bed composed of a very thick layer of extremely soft clay interbedded with thin sand lenses saturated with arte-

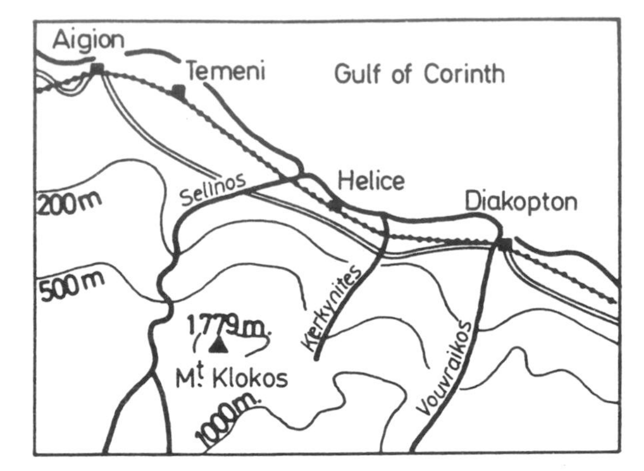

Figure 109. Location of Helice.

Figure 111. Location of Sybaris in Magna Graecia.

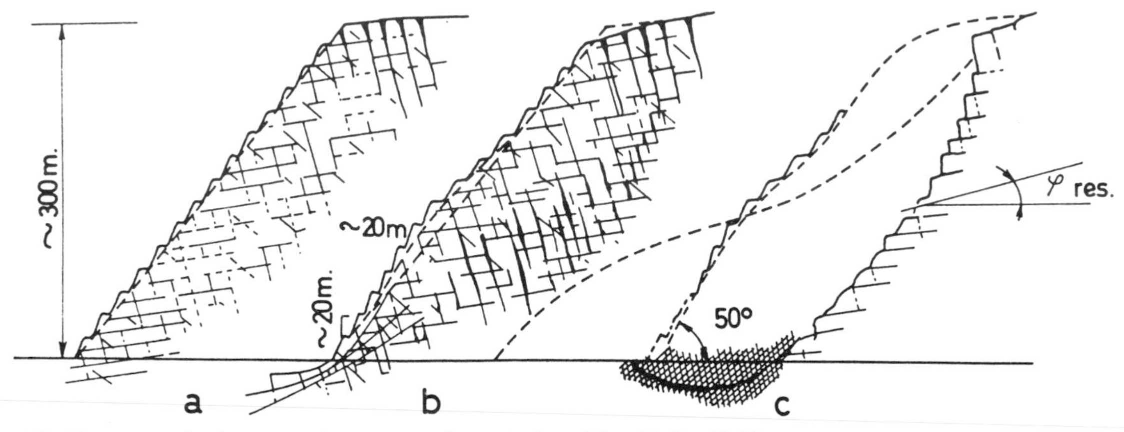

Figure 110. Three stages in the progressive rupture of a rock slope (after Muller, 1963):
a) Opening of stress fissures at the top of the slope;
b) Dislocation of the rock mass in the area of the rupture and concentration of stresses at the base;
c) Rupture at the base of the slope. The dashed line shows profile after the collapse.

sian water. In 1970, the author was asked to design a fresh project in co-operation with Professor Matteotti, an Italian specialist in seaports; the new scheme made extensive use of the Sumerian techniques we have described, adapted to modern materials. The port is now nearing completion and, so far at least, Poseidon has spared the modern child of Sibari from the fate of its ancient forebear.

MAN'S FORMER ATTITUDE TO EARTHQUAKES

The Homeric poems of the eight century B.C., and many other writings of the Greeks, contain descriptions of earthquakes showing that the Greeks were both superstitious and rational: superstitious in their belief in Poseidon, the god of soil and rocks, who could make the earth tremble and shatter the land, but rational in their observation of warning signs such as the behaviour of animals and the diversion of springs.

It will be remembered from an earlier chapter that it was the Greeks who invented the first anti-vibration mats and made cramps to prevent the joints of their stonework from breaking. Judging from the account of the seige of Apollonia, they had also discovered that the soil transmits shock waves. After the Greeks, however, there seems to have been a resigned attitude to these natural hazards that lasted for over a thousand years.

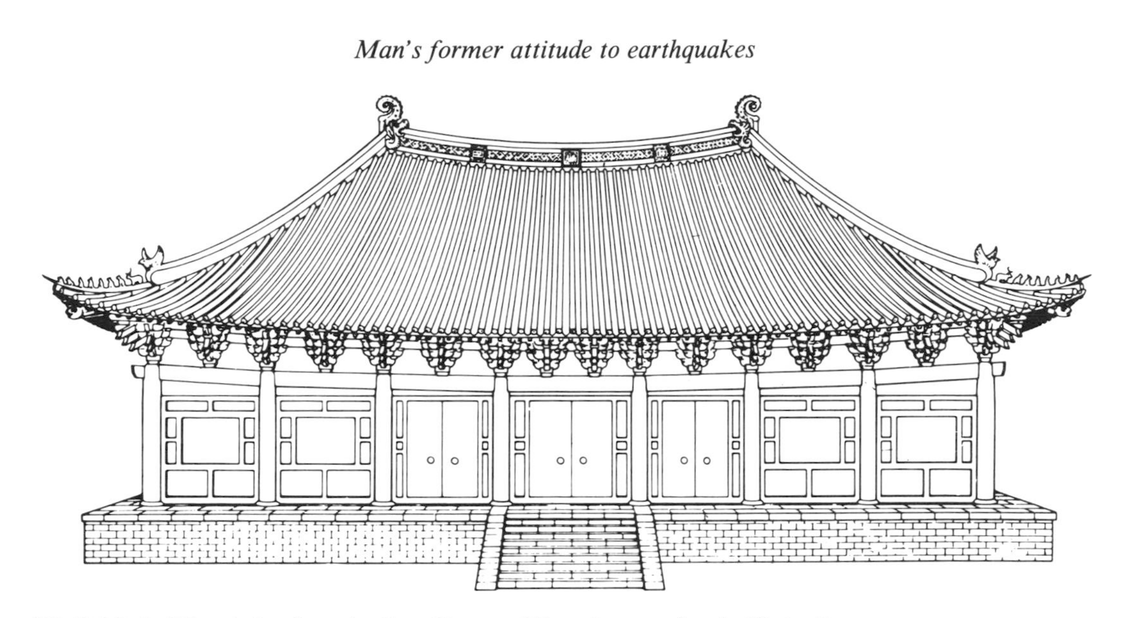

Figure 112. Public building dating from the Sung Dynasty (Eleventh century) and still standing.

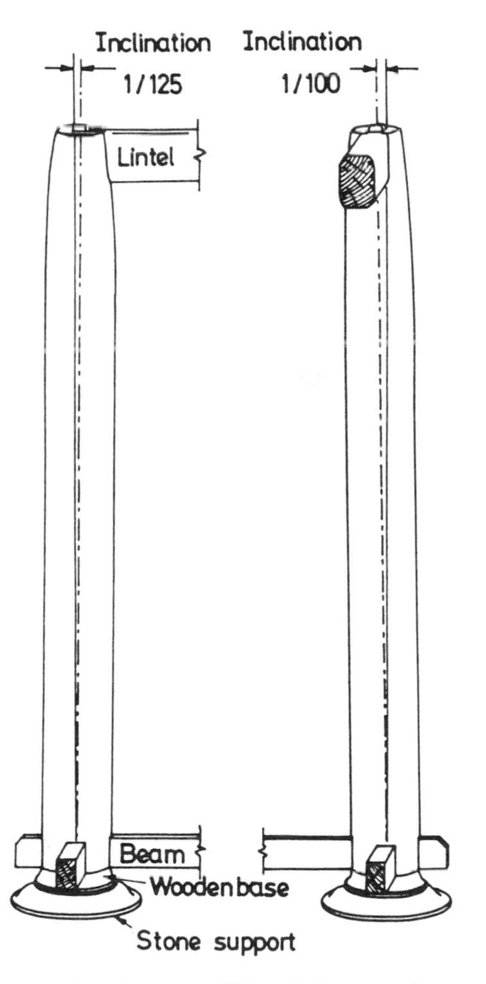

Figure 113. Eleventh century China. Columns resting on stone supports.

Tenth century China was incontestably the mother of what we now call earthquake engineering. Like Greece, this country is very exposed to earthquakes. In the present century alone, the Tangshan Earthquake of 1976 had 800000 victims and two others, in 1927 and 1920, left 200000 and 100000 dead – by far the most murderous natural disasters in modern times. It is therefore not surprising that the Sung Code of 1103 A.D. paid particular attention to seismic phenomena and some of the buildings designed to this Code are still standing (Figure 112).

The Sung Code imposed the use of wood, especially white cedar, which has an excellent tensile strength that allows the roofing to be sufficiently flexible to damp out the vibrations without rupturing: the building thus sways at a slower frequency than the earth and does not enter into resonance.

In their technique, as can be seen from Figure 113, the columns were supported on a wooden base resting on a stone plinth laid on a compacted soil. The columns were linked to each other by horizontal wooden beams at the base and by slotted lintels at the top. The overall stability of the structure was improved by inclining the columns slightly inwards (0.8% in a north-south direction and 1% in an east-west direction).

Figure 114 shows how the roof was supported by jointed brackets. The roofs were very heavy, in order to withstand typhoons, and strongly curved so as to diminish the uplift force due to wind and pro-

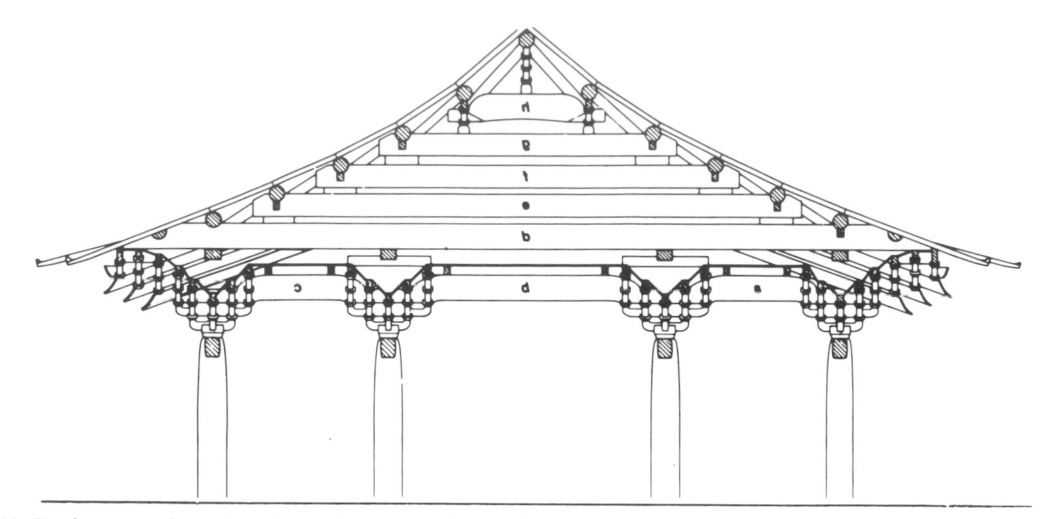

Figure 114. Roof supported on the columns by means of jointed brackets.

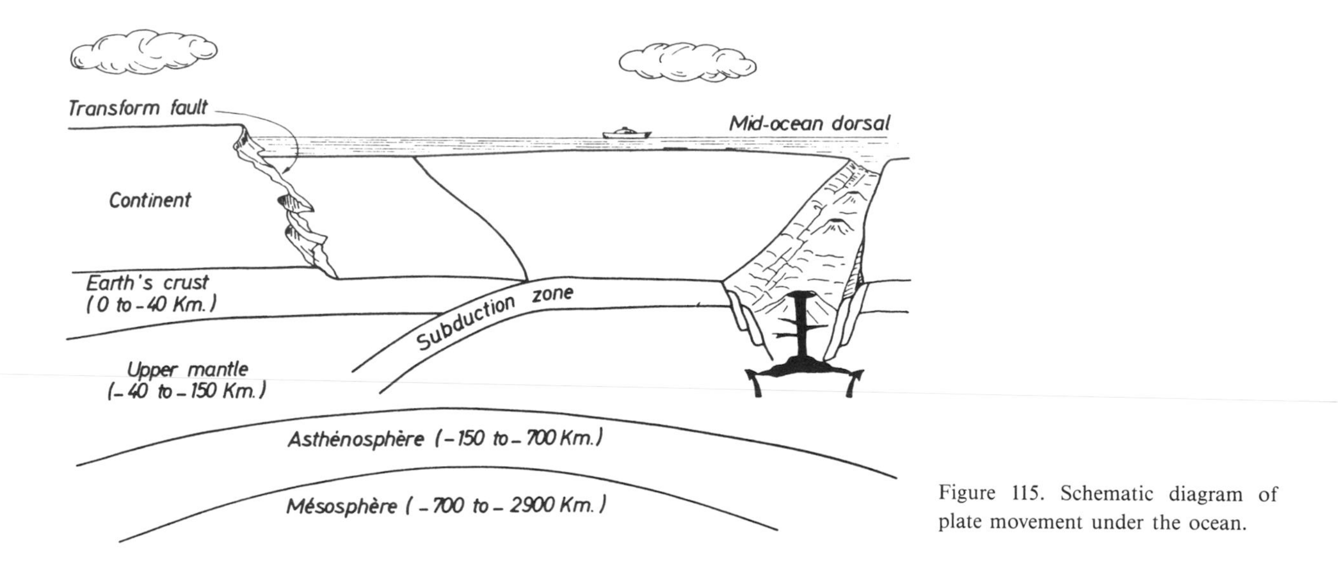

Figure 115. Schematic diagram of plate movement under the ocean.

mote rapid run-off during the heavy rains.

All in all, Chinese architecture shows how, by careful thought, they adapted their designs to satisfy certain functional criteria.

In the Middle Ages, and even later, earthquakes and the havoc they wrought were regarded as a punishment of God. The Fathers of the Church would have excommunicated anyone who dared to thwart the designs of the Creator.

The Lisbon Earthquake occurred in 1755, during the lifetime of Voltaire. In his 'Poem on the Lisbon disaster' he contradicts Pope's 'Essay on Man', a hymn to self-satisfied reason – 'Whatever is, is right' – and, while defending himself against the accusation of being anti-Christain, alludes, as Jean-Jacques Rousseau remarked, to a 'maleficent God whose only pleasure is to hurt'. This is an odd attitude on the part of Voltaire, not only because he was a man who denounced the obscurantism provoked by the excesses of religious zeal as 'Infamy' but because he was a writer who, long before, had drawn Europe's attention to the genius of Newton and thus helped to initiate the scientific era.

THE PRESENT ATTITUDE TO EARTHQUAKES

Never before have newspapers devoted so much space to earthquakes. They describe in detail the warning tremors that announce the coming dis-

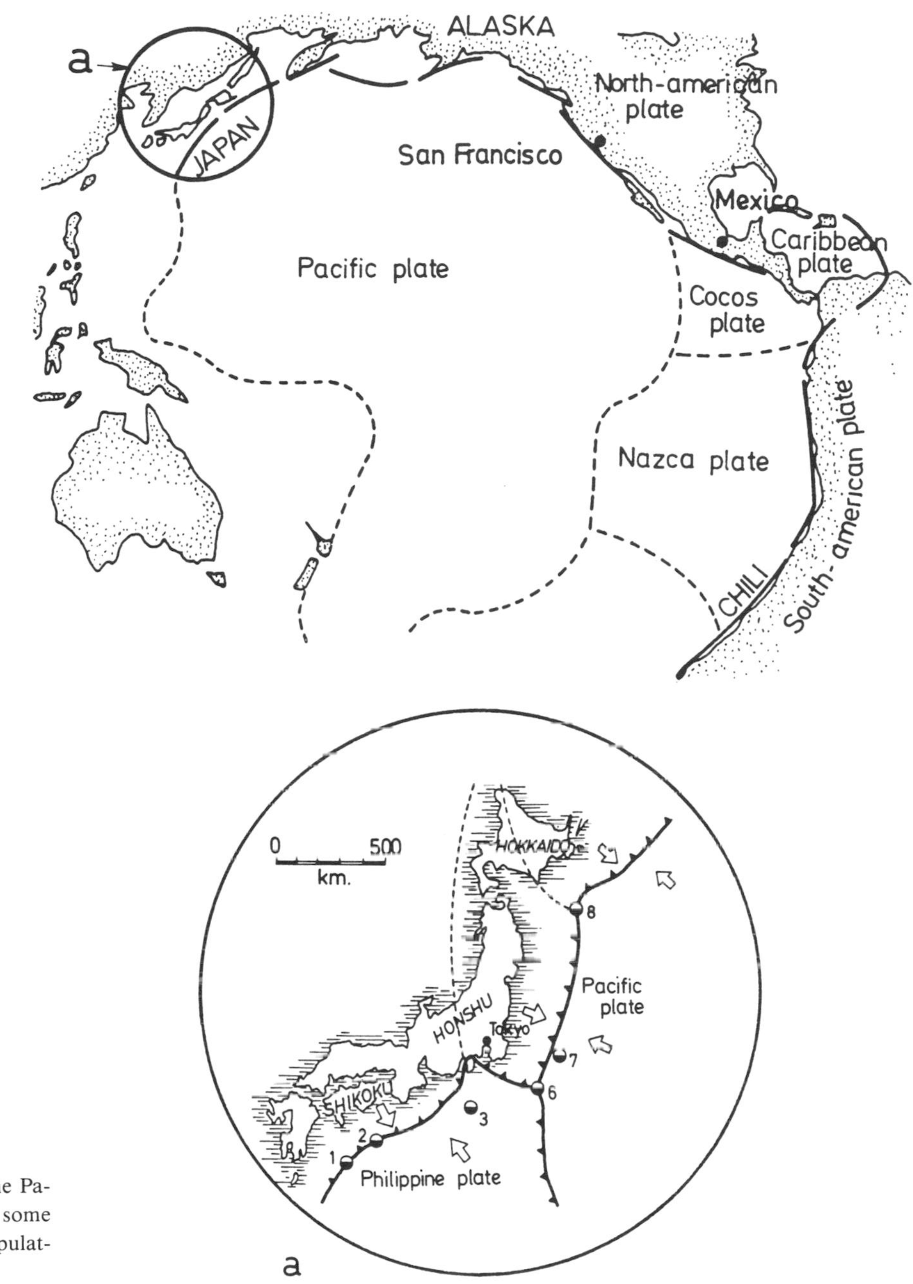

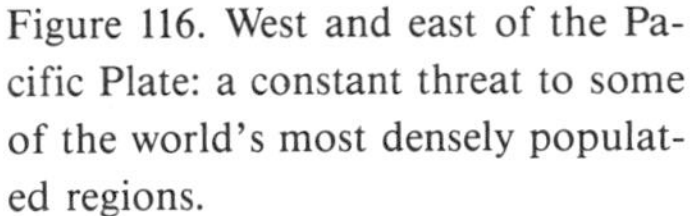
Figure 116. West and east of the Pacific Plate: a constant threat to some of the world's most densely populated regions.

aster, the low frequency compression waves that set off an ominous resonance, the shear waves that provoke displacements perpendicular to their paths and the ground waves that complete the disaster.

Many readers pay close attention to the description of the tectonic plates and microplates that float like rafts on the magma of the earth; they now know that these plates form the floor of the oceans and contain slowly widening fractures (the volcanic dorsals) through which the magma spews on to the ocean floor its molten material derived from other plates which, to make room for it, buckle and plunge beneath their neighbours (Figure 115). But the disappearance of the subducting plate (and of the sediments deposited upon it) does not take place smoothly and each lurch triggers an earthquake in the continental mass above.

Geologists have taken decades to digest their hostility to the ideas of Wegener on continental drift and his concept of plates floating on the magma. In contrast, the general public were immediately interested in and curious about this way of looking at things: it is not at all a matter of indifference to the average Mexican to know that the dangers threatening his country spring from the thrust of the Cocos Plate against it (Figure 116), and the Japanese are painfully aware that the plates of the Pacific and the Philippines are besieging their country, directing their battering-rams against the walls of the Eurasian Plate, with Tokyo one of the outposts of the defence.

But what forces drive these plates? And it is here that popular science is almost at the same level as the specialists, who seem unable to answer this fundamental question.

In the nautical games so beloved of Nero, it was possible to perceive at each instant the dominant force of the crew of gladiators of a quadrireme that was going to sink its opponent. Here we are up against a mystery.

Holmes, one of the first geologists to back the 'plate theory', attributed these forces to convection currents created by radioactivity (Figure 117) and although no actual figures have been produced this is still the prevailing idea. It is, therefore, not surprising that man has made little progress in the forecasting of earthquakes.

In China, children are still handed simple descriptions and drawings about the reactions of animals: 'When the earthquake approaches, fishes jump out of the water, hens cackle as if they had gone mad and perch in the trees; cows refuse to enter their sheds, dogs howl like whales, snakes come out of hibernation...'. But China, which has no wish to discard all the knowledge accumulated over several thousand years, marries the ancient and the modern by attending to the warnings of animals and at the same time processing the information obtained from seismographs and, in the near future, inclinometers (placed near the subduction trenches) which can react just before the onset of an earthquake.

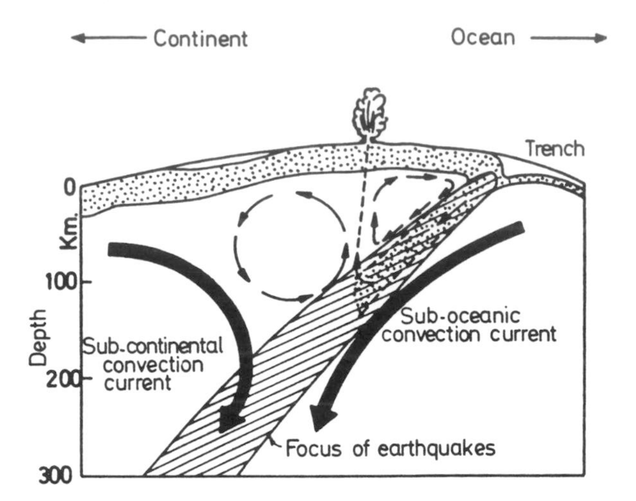

Figure 117. The diagram of Holmes illustrating the mechanism of subduction into the magma due to convection forces.

Figure 118. Chinese children are given simple drawings to teach them how to recognize an imminent earthquake: a horse refuses to enter its stable and pigs run away in panic.

All over the world, however, the highest hopes are being pinned on the progress of anti-seismic engineering. The question for us is whether man is making progress in the (invisible) art of building that will enable him to thwart the caprices of a soil unwilling to be tamed. Judging from the destruction caused by the most recent earthquakes, one might be tempted to say no – but that would be a superficial answer. In reality, undoubted progress is being made through meticulous analysis of the special wave characteristics and the damage caused by each earthquake, and the conclusions of this research are being used to steadily tighten up and improve anti-seismic regulations.

The result has been the publication of numerous building codes for the most endangered zones, but some of these are so detailed and complex that they not only raise costs but act as disincentives. The trouble is that, in the mass of regulations, a distinction is not always made between what is indispensable and what is secondary. All too often, the recommended methods of calculation disregard the real causes of collapse: it is considered enough to calculate the structure, which is assumed to be perfectly elastic, even though the elastic phase of a building is comfortably exceeded when it starts to rock. This has been demonstrated for frameworks of multiple reinforced concrete porticos: inadequately reinforced concrete breaks like glass whereas it becomes ductile and deforms like marshmallow when it has been properly reinforced. Rule 1 of the ABC of earthquake protection is to strengthen the concrete of vertical pillars with transversal reinforcing and to pay special attention to the intersection of beams and columns.

Lastly, and above all, it is becoming increasingly realized that earthquakes of the same magnitude can affect structures in very different ways owing to differences in the frequency of the vibrations, in the total duration of the earthquake, in the nature of the soil, and so on. The Mexico City earthquake provides a good illustration.

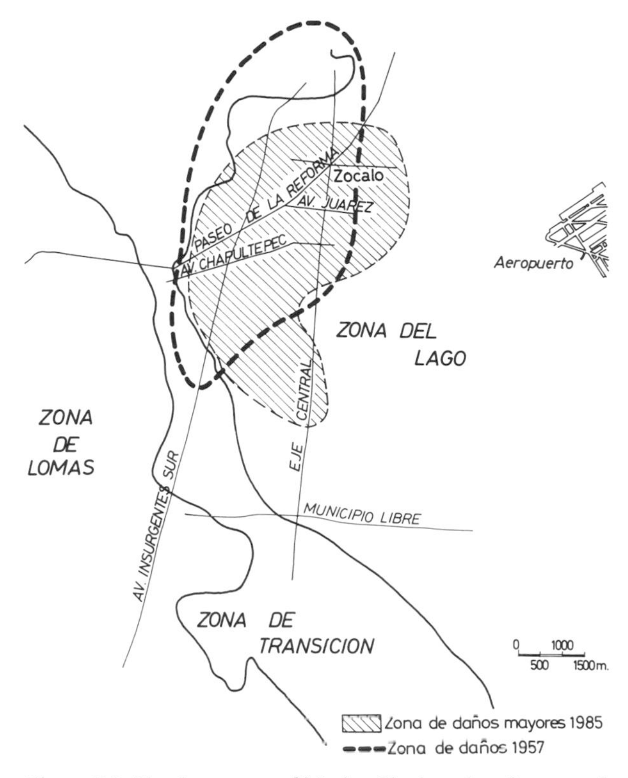

Figure 119. The three zones of Mexico City based on features of the subsoil.

THE MEXICO CITY EARTHQUAKE OF 19 SEPTEMBER 1985

Over 9000 dead, 50000 buildings damaged and 500 collapsed, 5-10 thousand million dollars worth of damage – all this caused by an earthquake of magnitude 8.1 and its violent aftershock on the following day.

And yet the giant capital of Mexico was a long way – over 350 km – from the epicentre, which was close to the subduction zone of the Cocos Plate (Figure 116). The earthquake was a singular event not only because of this long distance but also because of certain peculiarities related to the soil and underground topography of the city.

As we have seen, the old part of the city was founded on what used to be Lake Texcoco, on an artificial filling overlying clays saturated with water. The city has since expanded westward, towards a transition zone of firmer rock that corresponds to an ancient now buried cliff bordering the lake; there was less damage in this area and very little damage at all in the 'lomas', an undulating region further west, the soil of which is of good quality. In contrast, there was heavy damage in that part of the infilled Lake Texcoco closest to the underground cliff at the edge of the transition zone (Figure 119).

The accelerations in the ground were not enormous since they did not exceed 20% of that of gravity, a modest figure for an earthquake of magnitude 8.1 (to be compared with the 50% measured at San Fernando with a magnitude of 6.6), but the to-and-fro pulses were slow, with a two-second cycle as against the cycle of $\frac{1}{5}$ second at San Fernando. As a result the ground swayed back and forth by not less than 20 cm and, to make matters worse, the cycle was repeated twenty-three times. A sort of long groundswell was set in motion by the collision of these very unusual clays against the solid cliff, just as the oscillations of a seaswell double in amplitude when it strikes against a dyke.

Even if Mexico City's anti-seismic regulations had been scrupulously respected, which was not the case, they would not have worked for this particular 'temblor'. The ones on which they were based had not produced such unhurried and repeated pulses, generating a series of slow to-and-fro movements with an amplitude that destroyed those buildings unable to flex sufficiently. Many of the ruins showed clearly that, in the concrete-framed buildings, the transverse reinforcing of the pillars was inadequate or even absent and that the joints between beams and columns were too weak.

The effect of this particular earthquake was due to the nature of the soil. Oddly enough, however, only a few metres below the surface, that same soil provided the surest refuge for the inhabitants: there were no deaths, and no one injured, in the Underground. Moreover, while some 7000 leaks were counted in the water pipes near the surface, the drainage channels at a depth of 50 m suffered no damage at all.

It was also curious that, except in cases where no safety margin had been left, the foundations of buildings held firm. In Mexico City during the first half of this century, cunning - and wealthy - builders did not trust the clay but founded their constructions at a great depth by driving piles down to find support on the underlying sand. As mentioned in an earlier chapter, however, the clay under the city has settled by 5 m since the beginning of the century with the result that the buildings with a firm foundation on the deep sand saw their cellars transformed into the ground floor. So the Mexicans then tried using shorter piles, called floating piles, which are embedded in the structure of the clays and follow the clay down as it continues to settle. Knowing that earthquakes liquefy loose saturated sands, it could have been feared that the same thing would happen to these clays with their extremely high water content when they were shaken by the numerous cycles of deformation. But nothing occurred. Their structure did not change and they continued to support the piles.

Naturally, when tall buildings resting on piles (Photo 47) founded without a safety margin of resistance were subjected to the rocking motion that overloaded first one row of piles and then another, they reacted by leaning over, pushing down the piles on one side and tearing up those on the other. The building in Photo 48 reminds one of a tree that has been upended by a squall to reveal the number and depth of its roots. Other buildings that were not so high but whose foundations also lacked a safety margin simply sank down in one piece, reducing considerably the height of the ground floor (Photos 49 and 50). The building in Photo 50 has even dragged a part of the street pavement down with it so that its erstwhile ground floor has now become a cellar.

In considering the 1985 earthquake of Mexico City, one might be tempted to conclude that the effectiveness and future of anti-seismic engineering were not very promising. This would be a mistake. In the first place, it would be overlooking the fact that some other very tall modern buildings (such as the Latin American Tower) suffered no damage at all. Secondly, the building code was inadequate: the earthquake of 1957, which had also caused extensive damage in the same part of the city (Figure

Photo 47. A tall building that has overturned, uprooting its foundation piles. (Photo by Professor Canba).

Photo 48. A building that has toppled over backwards because the bearing capacity of its foundation piles on that side had an insufficient safety margin. (Photo by Professor Canba)

Photo 49. A building that has settled vertically in one piece. (Photo by Professor Canba)

Photo 50. Vertical settlement of a building together with part of the street. (Photo by Professor Canba)

119), had in fact been badly interpreted for the simple reason that no accelerograph had functioned. There is no doubt that future regulations requiring high levels of ductility in medium to high buildings will be much more effective. Even though the ground, which entered into resonance along the rim of the underground cliff, had made the structures built upon it lurch violently, at least it proved that it was still able to support them – a fact of the greatest significance for the future of this megalopolis.

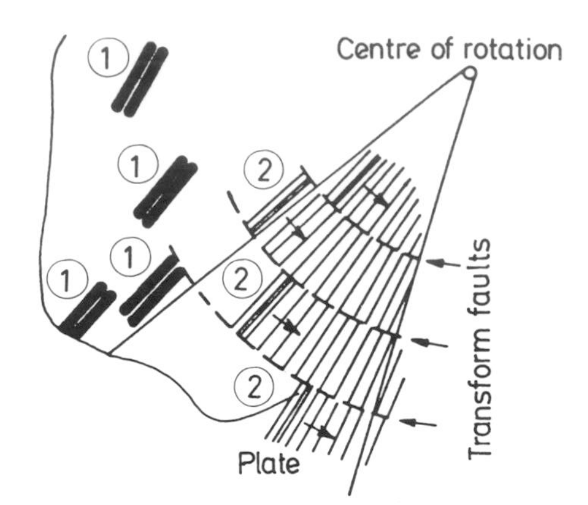

Figure 120. Circular transform faults in a dorsal which allow the plates to pivot.
1. Dorsal before pivoting
2. Dorsal after pivoting.

BUILDING ON THE EDGE OF A FAULT: THE SAN ANDREAS FAULT

The great ridges that indicate the movement of the tectonic plates are more striking on the ocean bed than on the continents. At the bottom of the oceans it is possible to trace the circular transform faults which, with their staggered dorsals, allow the plates to pivot (Figure 120). On the continents, by contrast, there are few continuous furrows over long distances: for example, it is simply a broken line that, for 4000 km from Mozambique to the Dead Sea via Lake Rudolph, Ethiopia, the Red Sea and the Gulf of Aqaba, signals the East African Rift System, a tear in the earth that poses a remote threat to the cradle of the ancient kingdoms of Judah and Israel and to the superimposed ruins of Jericho, nearly ten thousand years old... which will eventually suffer the same fate as Sodom (Genesis 19). By contrast, the San Andreas Fault forms an almost continuous line on the ground and constitutes a direct threat to Los Angeles, San Francisco and the booming part of California.

This fault attracted world attention at the time of the great earthquake and fire which ravaged San Francisco on 18 April 1906, killing 700 people. Yet it was not the first time it had woken up: the earthquake of 9 January 1857 had the same magnitude but fewer victims as California then had a much smaller population.

Of particular interest to us are the ways and means now employed by the inhabitants of the region to limit the damage in the highly industrialized region traversed by the fault.

The San Andreas fault is probably 100 million years old and is the 'master' fault (Figure 121) of an intricate network of faults (Figure 122) cutting through the coastal region of California. The main

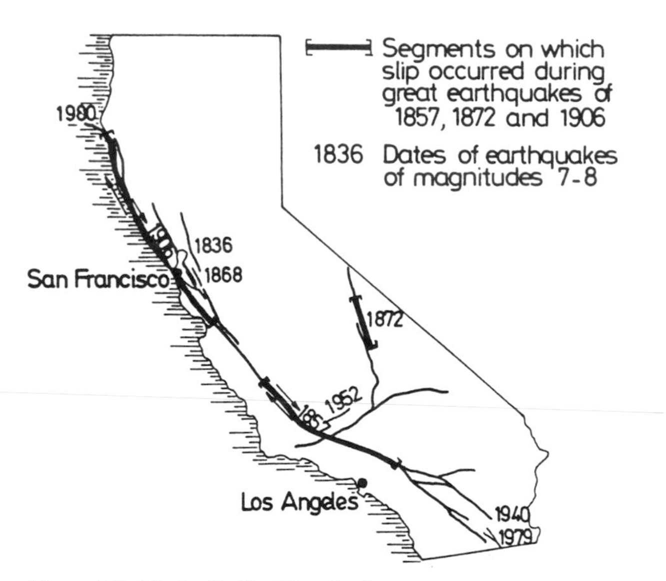

Figure 121. Master fault of San Andreas

fracture, some 1000 km long and 30 to 35 km deep, menaces both San Francisco and Los Angeles. It begins in the south under the Gulf of California and further north divides into two branches, the San Andreas Fault itself, threatening San Francisco, and the Hayward Fault threatening Oakland and Berkeley (Figure 72).

It is one of the rare visible shear faults that is developing quickly. The shearing action operates towards the right (Figure 123), i.e. if one were to stand on one side of the fault and look across it, the block opposite would appear to have moved to the right.

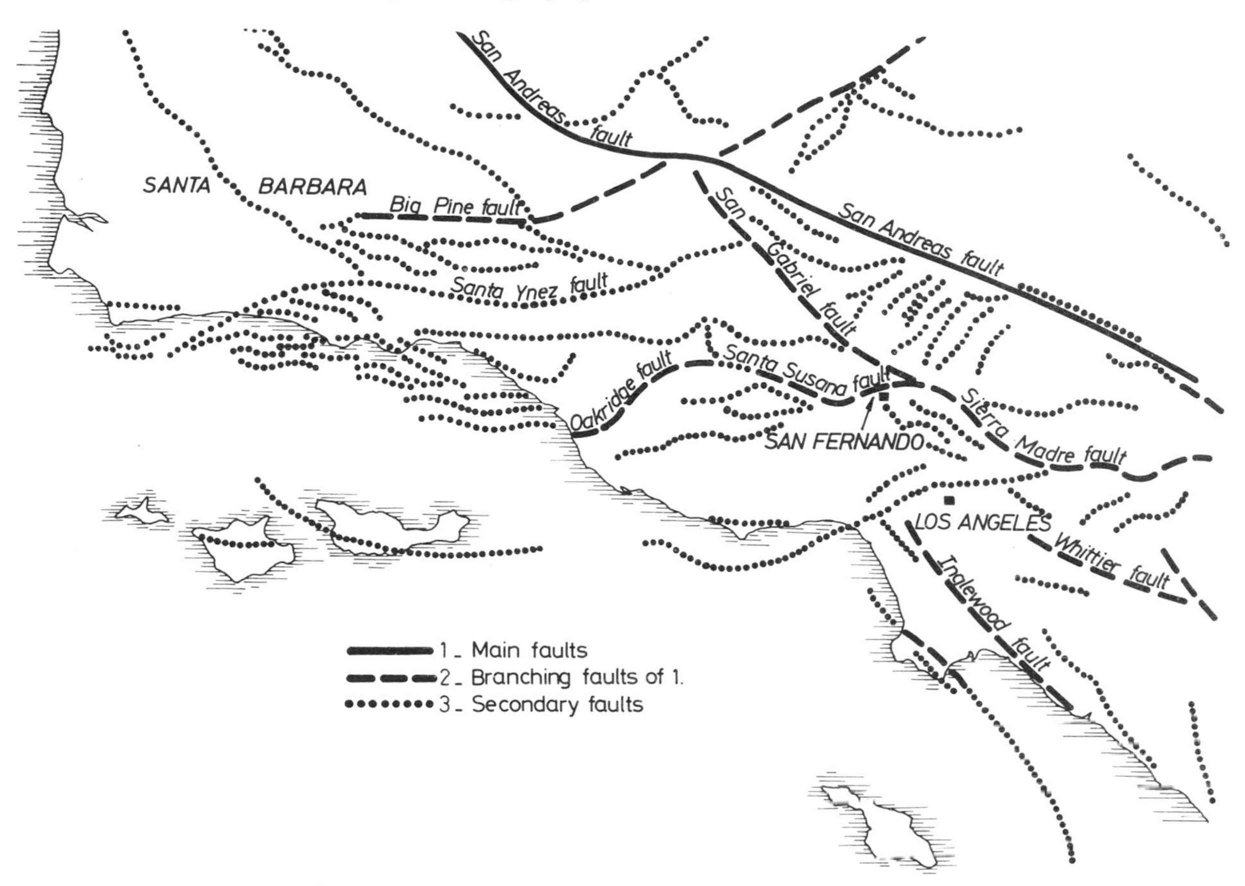

Figure 122. Intricate network of faults in the Los Angeles area. Note location of the San Fernando dam.

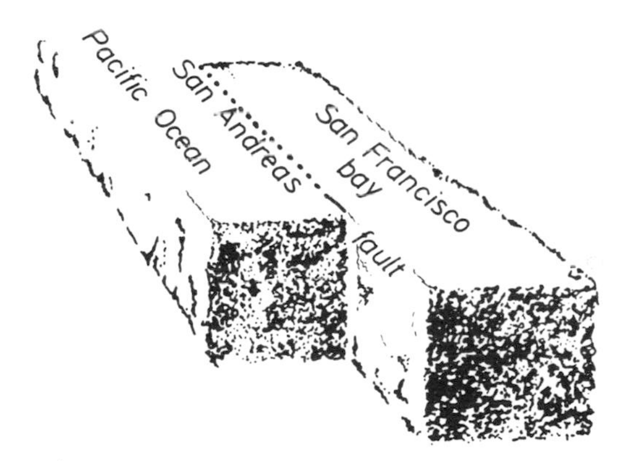

Figure 123. The slip movement along the San Andreas fault.

Geologists estimate that the total accumulated displacement along the fault now amounts to between 500 and 600 km. During the 1906 San Francisco earthquake alone, a sideways wrench of 6.30 m was measured.

Thus the Pacific Plate is planing California in a north-west direction and will eventually slice off the richest part of the State from the North American continent.

At present, the lateral drift averages about 5 cm per year (measured about 30 km south of San Francisco). But the slip movement is not uniform throughout the length of the fault; on the contrary, the measurements taken reveal that, all along the fault, sections that are sliding alternate with others where the movement is blocked – not necessarily a good thing since a great deal of energy may be accumulating there. One of the latter spots in particular lies under the urban area of San Francisco.

The morphology of the fault also varies from place to place. In the Carrizo Plains, it takes the form of a distinct but fairly narrow trough; about 30 km south of San Francisco (Photo 51) it carries an elongated lake flanked by rising ground. In other spots, however, such as at Hayward to the south of Berkeley, the strike-slip affects a much wider area within which, as the years go by, the roads and

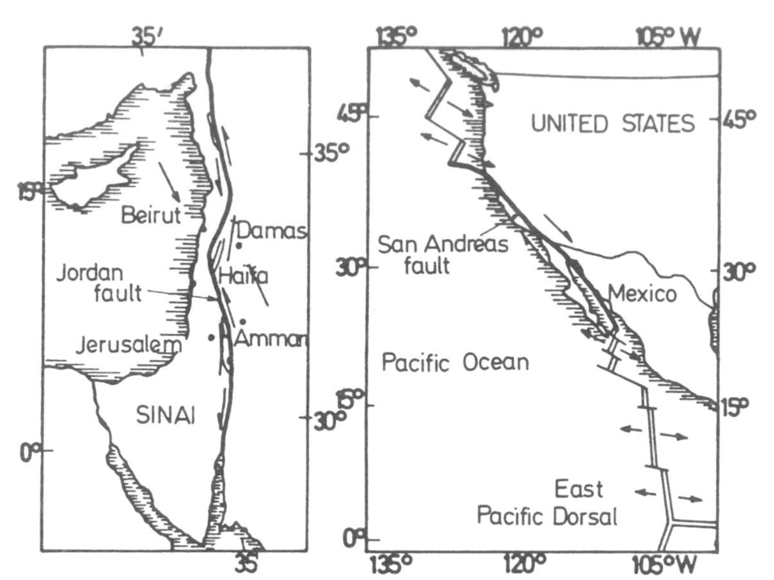

Figure 124. Two slip faults
. The Jordan fault: a left-lateral or sinistral fault
. The San Andreas fault: a right-lateral or dextral fault

Photo 51. A view of the San Andreas Fault. Near San Francisco.

pavements slowly bend and certain well-braced buildings pivot slightly without suffering any damage.

In California, the US Geological Survey has 300 geologists keeping a weather eye on the morphology of the fault and the seismicity of the region. Lasers are being used to obtain extremely precise measurements of the movement, and highly sensitive listening devices pinpoint the daily epicentres of the tremors with a magnitude of less than 2, inaudible to human beings.

The records of such microshocks as well as the moderate and major earthquakes make it possible, by extrapolation, to calculate with reasonable accuracy the average period of recurrence of earthquakes with a given magnitude. The ones in 1857 and 1906 had a magnitude of 8.3 while that of 1971 did not exceed 6.6, but we have seen the disastrous effect it had on certain dams in the neighbourhood of the fault.

In short, the period of recurrence of an earthquake of magnitude 8.3 is estimated at 200 years

Photo 52. Housing area near San Francisco straddling the San Andreas fault. (Photo by Robert E. Wallace, U.S. Geological Survey)

(on average). In other words, by the year 14000, i.e. as far in the future as the first human constructions reach back into the past, the city of San Francisco will have been struck by 12000 / 200 = 60 earthquakes of this magnitude.

Strangely, however, Californian town planning does not always appear to take this latent danger fully into account. Robert Wallace, in a document issued by the United States Geological Survey, states that 'Geologists are horrified to see land developers build rows of houses straddling the trace of the 1906 break' (Photo 52).

In the United States, the municipalities have full control over the regulation of town planning. In San Francisco, the tallest building is 237 m high and has 52 storeys; though the municipality now opposes the construction of further skyscrapers, its decision owes more to the general trend to 'de-Manhattanize' now spreading across the whole country than to anti-earthquake measures.

Three factors, however, offer some reassurance for the future of this highly urbanized zone. In the first place, the numerous observations of earthquakes in the region (accelerations, frequencies, duration, etc). suggest that, unlike the case with Mexico City, the anti-seismic regulations will measure up to future earthquakes; secondly, it is unthinkable that this building code will be violated since the danger is so immediate; thirdly, anti-seismic techniques are steadily improving: for instance, certain buildings in particularly exposed areas, whose collapse would be catastrophic (nuclear power stations), are now being founded on elastomer pads reinforced with steel plates which can support heavy loads and at the same time withstand lateral displacements. With foundations such as these, the building behaves like an oscillator with a period of resonance that can be 'tuned' by the builder to vibrate at a frequency far removed from that of the earthquake. In consequence, the bold but carefully considered architecture of big buildings located on the line or edge of the fault are likely to be punished much less severely than the hastily conceived housing projects of certain developers in the suburbs (Photo 52).

'California, with all thy faults, we love thee still' is an amusing play on words with a blackish tinge to its humour. California is in the midst of a geological process of transformation. It is, as John Muir once said, 'living in creation's dawn', yet it is on this land that one of the most dynamic of civilizations is leaving its imprint. Fully aware of the dangers, its architects and engineers are striving to anticipate them by careful study of the most serious risks.

Conclusion

The myths connected with the earth – the earth as mother, the earth as nurse, the earth as lover – have often been depicted in literature, painting and sculpture. Our concern has been to regard her as a storehouse of memories, as the custodian of man's imprints and the bearer of his creations, an earth that allows herself to be hollowed, cleaved and opened as man penetrates her surface, penetrates her depths and even the bottom of her seas.

She keeps a record of man's passage marked by a host of master-works but also by periods of oblivion followed by fresh starts after man's rediscovery of forgotten skills. The earth's crust is like a book whose pages recount the lessons of our past, though some of these pages have been lost because the writing has slowly faded, because the leaves have been brutally torn out or because builders have scorned to read them.

The visible marks on her surface do not tell the whole story: there also exists a world of concealed forms – and we have sought to give the reader a few glimpses of the mysterious hidden face of architecture and kindle his interest in man's invisible 'geography' – his 'writing in the soil' – which is still largely undeciphered.

She is an earth who responds to gentleness in the art of builders, who is willing to sustain their creations and aid their skyward ambitions, but who is also fickle and ready to topple the works that demand her assistance without grace.

She can be fearsome too, when she liberates the forces straining against her crust and shakes like reeds the tall edifices raised by man, sparing those which have understood the mysteries of her agitation but pitiless for the others.

Oscar Wilde, as severe for doctors as he was for architects, said that 'doctors bury their mistakes but not architects!' But foundation engineers bury both their masterpieces and their mistakes; like the nameless architect of the Tower of Pisa, they are hardly known other than through the unexpected consequences of their errors. In this book we have tried to rehabilitate their work. As Fulgence Bienvenue, the 'Father of the Paris Metro' wrote: 'It is the characteristic of underground constructions to show no gratitude to their authors. The truth is that, when the difficulties are completely overcome, the finished work should show no trace of them and the men who have achieved this victory are obliged to ask people to trust their word'.

When Coulomb presented his celebrated Memoir to the Royal Academy of Science in 1773, he was even more modest: 'While great men, installed in the roof of the building, design and build the upper storeys, ordinary workmen, scattered in the lower storeys, or hidden in the darkness of the foundations, should try to perfect that which more capable hands have created'.

We would be wrong to think that the future will not obliterate some of the major tokens of our rapidly expanding civilization; within a time-span no longer than that which separates us from the first habitations, a number of these proud symbols will have gone the way of Helice or the forest of Scissay.

'Tu es Petrus et super hanc petram...' – The Lord tells us to build upon stone, upon rock: 'And the rain descended, and the floods came, and the winds blew, and beat upon that house; and it fell not, for it was founded upon a rock'. The Church has taken this image to mark its permanence and its invisible strength. 'Super hanc petram...' – 'Upon this rock...'. In the works of man, too, the foundation is essential. The men who knew this did not build only upon rock but in every place where civilizations developed and, when all is said, our mother earth has often looked kindly on their efforts and accepted the burden of their constructions. It is their successes and their failures in the invisible art of this relationship that we have attempted to describe.

ACKNOWLEDGEMENT

I owe a profound debt of gratitude to Philip Cockle who has, throughout these pages, not only contributed greatly to their expression in English but has also offered valuable suggestions and criticism on various parts of the manuscript.

Bibliography

INTRODUCTION AND PART ONE

Alain, E. 1958. *Les Arts et les dieux.* Paris: Encycl. La Pléiade: 567.

Ambraseys, N. N. 1975. *Engineering Seismology.* Inaugural lecture, University of London.

Bonaparte, Colbert, Currie, de Ricqles, Kielan-Jawozowska, Leonardi, Morello & Taquet 1984. *Sulle orme dei dinosauri.* Venice: Erizzo.

Borchardt, L. 1911. *Die Pyramiden, ihre Enstehung und Entwicklung.* Berlin.

Caquot, A. in Kerisel, J. 1978. *Albert Caquot (1881-1976), Créateur et précurseur.* Paris: Eyrolles

Carillo, N. 1948. Influence of Artesian Wells in the Sinking of Mexico City. *Proc. 2nd Int. Conf. Soil Mech. Fdn. Engng.* Rotterdam: Kiesmaat.

Clarke, & Engelbach, R. 1930. *Ancient Egyptian Masonry.* Oxford.

Coles, B. & J. 1986. *Sweet track to Glastonbury : The somerset levels in Prehistory* . London: Thames and Hudson.

Conant, K. J. 1968. *Cluny.* Macon: Protat.

Croce, A. 1985. Old monuments and cities. Research and preservation. *Geotechnical Engineering in Italy. Associazione Geot. Italiana.* 362.415

Creazza, G. 1984. Constructions monumentales dans le milieu lagunaire à Venise. *Symposium de Sciences Humaines. Université de Naples.*

Danyzy, A. A. H. 1732. Méthode générale pour déterminer la résistance qu'il faut opposer à la poussée des voûtes, *in Hist. de la Soc. Roy. des Sciences établie à Montpellier.* Lyon: (1778): 2-40

Delétie P., J. Montluçon, J. Lakshmanan & Y. Lemoine, (in press). *Aspects techniques et physiques de l'operation Kheops.* Ann. Inst. Techn. Bat.

Derchain, Ph. 1969. Snefrou et les rameuses: commentaire sur le papyrus Westcar, *Rev. d'Egyptologie:* 19-25.

Edwards, I. E. S. 1982. *The Pyramids of Egypt.* Harmondsworth: Penguin.

Emery, W. B. 1949–58. *Great Tombs of the 1st Dynasty,* 3 vols., Cairo and London.

Fernandez, C. 1961. Las presas romanas en Espana, *Revt. ob. pùbl.:* 357-363.

Ginouvès, R. 1966. Note sur quelques relations numériques dans la construction des fondations des temples grecs. *B.C.H.:* 104-107.

Gullini, G., R. Parapetti & G. Chiari 1985. *La terra tra i due fiumi.* Torino: Il Quadrante Edizioni: 241-259.

Hawkes. 1974. *Atlas of Ancient Archeology.* London.

Heyman, J. 1969. The Safety of Masonry Arches, *Int. J. mech. sci.:* 11, 363.

Hu Taho, Xia Schulin et al. 1983. An Investigation on the Foundation of the Ancien An-Chi Bridge (in Chinese), *a paper submitted to the 1st Conf. on the History of China's Science.* March: Kunming

Ishihara, K. et al. 1977. Blast Furnace Foundations in Japan, *IXth Int. Conf. Soil Mech. Fdn. Engng., Spec. lect.* Tokyo: 157-236.

Kerisel, J. 1956. Historique de la mécanique des sols en France jusqu'au 20e siècle. *Géotechnique.* Dec: 151-166.

Kerisel, J. 1973. Le barrage d'Arzal: un barrage sur sol très compressible construit au travers d'un estuaire à marée. *Géotechnique.* March: N° 1, 23, 49-65.

Kerisel, J. 1975. Old Structures in Relation to Soil Conditions: *15th Rankine lecture. Geotechnique:* London: 25. N°3: 433-483.

Kerisel, J. 1982. Les chantiers et monuments du passé (le Panthéon français). *Ann. Inst. Techn. Bat.* 410: 55-60.

Kerisel, J. 1985. The History of Geotechnical Engineering up until 1700. *Golden Jubilee Book. XIth Conf. Soil Mech. Fdn. Engng.* San Francisco. Rotterdam: Balkema.

Koldewey, R. 1913. *Das weider erstehende Babylon.* Leipzig. *In Engl. transl. by A. S. Johns 1914. The excavations at Babylon.* London.

Lauer, J. P. 1974. *Le mystère des pyramides.* Paris: Presses de la Cité.

Leakey, M. 1979. 3,6 Million Years Old Footprints in the Ashes of Time. *Nat. Geogr. Mag. April:* 446-457.

Lees, G. N. & N. L. Falcon. 1952. The Geographical History of the Mesopotamian Plains. *The Geographical Journal.* 118.

Lloyd, S. 1978. *The Archaeology of Mesopotamia.* London: Thames and Hudson.

Manasara Shilpashastra (Architecture) in Engl. transl. by Allen G. & Unwin. 1964. *The Eastern Key.* London.

Marogioglio, V. & C. Rinaldi. 1964. *L'architectura delle piramidi Menfite.* III. Rapallo: Artigrafiche Canessa.

Matos Moctezuma, E. 1982. El Templo Mayor de Tenochtitlan. Planos Cortes y perspectivas, *Instituto Nacional de Antropologia e Historia.* Mexico. D.F.

Mazari, M., R.J. Marsal & J. Alberro. 1985. The Settlements of the Aztec Great Temple, *Contr. to the XIth Conf. Soil Mech.*

Fdn. Engng. San Francisco. Rotterdam: Balkema.

Mendelssohn, K. 1974. *The riddle of the Pyramids.* N. Y.: Praeger.

Mooney, J. 1893. The Ghost-Dance Religion and the Sioux Outbreak of 1890. *Annual report f 21 of the Bureau of American Ethnology.* XIV, 2: 72. Washington.

Regourd, M., J. Kerisel & P. Delétie (in press). Microstructure of mortars from, Cheops Meidum and Unas Pyramids (respectively 2630 BC, 2500 BC and 2350 BC). *Cement and Concrete Research.*

Rondelet, J. 1804. Mémoire historique sur le dôme du Panthéon français. *Bibl. hist. Ville de Paris.*

Russo, F. 1983. *Nature et méthode de l'histoire des sciences.* Paris: Blanchard.

Russo, F. 1986. *Introduction à l'Histoire des Techniques.* Paris: Blanchard.

Safar, F. 1950. Eridu. A preliminary Report on the Third Season's Excavations, *Sumer,* 6.

Salvatori, N. 1984. *L'Uomo di Isermia, un italiano di 700,000 anni fa.* Milano Airone IV. N° 40: 79-101.

Schliemann, H. 1875. *Troie et ses ruines.*

Vauban (de) 1689. *Oisiveté ou Ramas de plusieurs Mémoires de sa façon sur différents sujets.* 12 Chap.: VIII Attaque des Places – IX Défense des places – XI Instructions sur le remuement des terres.

Viollet le Duc, E. E. 1866. *Article sur les restaurations du point de vue des fondations.* Dictionnaire de l'Architecture. Paris: Morel.

Vitruve (ler siècle avant J.C.) De Re Architectura, in Perrault 1684. *Les dix livres d'Architecture.* Paris: Coignard. English transl. 1955. Cambridge Mass: Harper and London: Heineman.

Vaughan, P. 1983. La fonction des outils préhistoriques. *La recherche.* 148 Vol. 14.

Wainwright, G. A., W. M. F. Petrie & E. MacKay. 1910 *Meydum and Memphis III.* London.

Wildung, D. 1969. *Die Rolle Agyptischer Könige im bewustein.* Berlin: Verlag Bruno Hessling.

Woolley, C. L. 1939. *Ur excavations. The Ziggurat and its surroundings.* London and Philadelphia.

Wu Rukang & Ling Shenglong. 1982. L'homme de Pékin. *Pour la science.* (french ed. of the Sci. Amer.) Août: 21-29.

PART TWO

Anguiloza, G. 1980. *Philippe Bunau Varilla: the Man Behind the Canal.* Chicago: Nelson Hall.

Atterberg, A. M. 1911. *Lerornas fohallande till vaten, deras plasticitetsgranser och plasticitetsgrader. Landtbruke Akademiens Handlingar och Tidskrift.* 50: 132-158.

Bauer, G. 1556 (Agricola G. 1494-1555) *De Re Metallica.* Bâle. *Engl. transl. 1919.* London Reed.

Bérigny, Ch. 1832. Mémoire sur un procédé d'injection propre à prévenir ou arrêter les filtrations sous les fondations des ouvrages hydrauliques. *Ann. Ponts & Chauss.* Août.

Boussinesq, J. 1892. Sur la détermination de l'épaisseur minimum que doit avoir un mur vertical, d'une hauteur et d'une densité données, pour contenir un massif terreux, sans cohésion, dont la surface supérieure est horizontale. *Ann. Ponts & Chauss.* 6th series 6: 494-510.

Boussinesq, J. 1884. Compléments à de précédentes notes sur la poussée des terres. *Ann. Ponts & Chauss.* 6th series 7: 443.

Boussinesq, J. 1885. *Application des potentiels à l'étude de l'équilibre et du mouvement des solides élastiques.* Nouvelle édition (1969). Paris: Blanchard.

Caquot, A. & Kerisel, J. 1948. *Tables de butée, de poussée et de force portante des fondations. English transl. by M. A. Bec. 1948.* Tables of the Calculation of Passive Pressure and Active Pressure. Paris: Gauthier Villars. 2nd bilingual edit. with Absi. (1973). Paris: Gauthier Villars.

Collin, A. 1846. *Recherches expérimentales sur les glissements spontanés des terrains argileux.* Paris: (Transl. by Schriever, W. R. 1955). *Landslides in clays.* Toronto).

Coulomb, C. A. 1773. Essai sur une application des règles de maximis et minimis à quelques problèmes de statique relatifs à l'architecture. *Mémoires de mathématique et de physique présentés à l'Académie Royale des Sciences.* Paris: Vol. 7, 343-82.

Danyzy. See Chap. 1.

Darcy, H. P. G. 1856. *Les fontaines publiques de la Ville de Dijon.* Paris: Delmont.

Darwin, G. H. 1883. On the Horizontal Thrust of a Mass of Sand. *Proc. Inst. Civ. Eng.* 71: 350-378.

Dettwiller, J. 1969. Le vent au sommet de la Tour Eiffel. *Monographie de la Météorologie Nationale.* Paris, 64: 13.

Eiffel, G. 1890. *La Tour de 300 mètres.* Paris: Lemercier.

Frontard, J. 1914. Notice sur l'accident de la digue de Charmes. *Ann. Ponts & Chauss.* 9th Series, 23: 173-292.

Gaffard, J. 1654. *Le monde souterrain ou description historique et physique de tous les plus beaux antres ...*

Gévin, P. 1973. *C. R. Acd. Sci.* Paris, 276.

Gillmor, C. S. 1908. Charles Augustin Coulomb: Physics and Engineering in Eithteenth Century France. *Dissertation for degree of Ph. D. Princeton University,* University microfilm *N° 69.2741.*

Glossop, R. 1960 and 1961. The invention and development of Injection processes. *Géotechnique.* Part 1 – Sept. 1960. Part 2 – Dec. 1961.

Hellström, B. 1952. The Oldest Dam in the World. *Houille blanche:* 423-430.

Heyman, J. 1972. *Coulomb's Mémoir on Statics.* Cambridge: University Press.

Kerisel, J. 1956, 1975, 1982. See Chapter 1.

Kerisel, J. 1973. Bicentenaire de l'essai de 1773 de Charles Augustin Coulomb. *Sp. lect. Proc. VIIIth, Int. Conf. Soil Mech. Fdn. Engng.* Moscow.

Kozlovsky, Ye. 1984. The World's Deepest Well, *Sci. Amer.* Dec. 106.

McCullough, D. 1977. *The Path Between the Seas. The Creation of the Panama Canal:* 1870 – 1914. New York: Simon and Schuster.

Mainstone, R. J. 1981. *The Eddystone Lighthouse* in Skempton, A.W. 1981. *John Smeaton.* London: Telford: 83-102.

Moreau, M. 1832. Notice sur une nouvelle manière de construire en mauvais terrain. *Mémorial de l'Officier du Génie* N° 11.

Murray, G. W. 1947. A note on the El Khafara: the Ancient Dam in the Wadi Garawi, *Bull. Inst. Egypt.* T 28, Sess. 1945-46.

Peck, R. 1985. Last Sixty Years. *Golden Jubilee Volume XIth*

Conf. Soil Mech. Fdn. Engng.: 122-133. San Francisco. Rotterdam: Balkema.

Poleni, G. 1748. *Memorie istoriche della gran cupola del Tempio Vaticano,* Padova.

Reynolds, O. 1887. Experiments showing dilatancy, a property of granular material possibly connected with gravitation. *Proc. Roy. Inst. London* 11: 354-363.

Schnitter, N. J. 1979. *Roman Dams, Water Supply and Management.* Pergamon Press 3: 29-39.

Seed, H. B. 1979. Considerations in the earthquake-resistant design of earth and rockfill dams, *Rankine lecture. Géotechnique.* N° 3, 215-263.

Seed, H. B. 1981. Rupture of the Teton Dam. Special lecture, *Proc. Xth Conf. Soil Mech. Engng.* Stockholm. Rotterdam: Balkema.

Skempton, A. W. 1985. A history of Soil Properties, 1717-1927 *Golden Jubilee volume, XIth Conf. Soil Mech. Fdn. Engng.* San Francicso. Rotterdam: Balkema: 92-122.

Smeaton, J. A. 1791. *Narrative of the building and a description of the Contraction of the Eddystone Lighthouse.* London.

Sokolowskii, V. V. *Statics of Soil Media.* (translated by D.H. Jones & A.N. Schofield 1960) London.

Solecki, R. S. 1971 *Shanidar: the humanity of Neanderthal man* London.

Stevenson, D. A., 1979. *The world's lighthouses before 1820.* London: Oxford Univ. Press.

Tacite, Annales. Livre XII. 56-57 of Budé's french transl.

Terzaghi, K. 1925. *Erdbaumechanik auf bodenphisikalischer Grundlage.* Vienna: Deuticke.

Terzaghi, K. 1926. *Principles of Soil Mechnics.* New-York: MacGraw Hill

Verne, J. 1864. *Voyage au centre de la Terre.* Paris: Hachette.

Vitruve. See Chapter 1.

Wooley, C. L. 1933. Report on the excavations at Ur. *The Antiquaries Journal.*

PART THREE

Berke, K. E., B. Menke & H. Streil, 1979. The quaternary geoligical development of the German part of the North-Sea. *Acta Univ. Ups. Symp. Univ. Ups.* Uppsala.

Chimalpahin. XVIIth cent. de San Anton Munon) (written in Nahuatl, Spanish transl. by Randon S. 1965: Mexico). *Relaciones originales de Chalco Amaquamecan. Memorial breve acerca de la Ciudad de Culhuacan Mexico.* Folio 60. 54-55.

Curtis, G. & J. Rutherford. 1981. Expansive Shale Damage, Theban Royal Tombs. *Xth Int. Conf. Soil Mech. Fdn. Engng. Stockholm.* Rotterdam: Balkema

Fernandez, D. 1980. *Le promeneur amoureux : de Venise à Palerme.* Paris: Plon.

Ghetti, A. & M. Batisse. 1983. La protection d'ensemble de Venise, *Nature et Ressources, UNESCO* XIX (4): 7-19.

Gilbert, S. & R. Horner. 1984. *The Thames barrier.* London: Telford.

Guéry, G., P. Pirrazoli & P. Trousset. 1981. Les carrières littorales de la Couronne. Indices de variation du niveau marin, *Histoire et Archéologie: Ports et Villes englouties,* Ser. 50.

Gullini, G., R. Parapetti & G. Chiari. 1985. *Attivita di consulenza, progrettazione et collaborazione scientifica : La terra tra i due fiumi.* Alessandria: Il Quadrante: 241-253

Haarnegel, W. *Die Grabung Feddersen Wierde,* Feddersen Wierde 2, Wieshaden. Steiner-Verlag.

Heraclitus. See Burnet, J. 1930. *Early Greek Philosophy.* London: A. & C. Clark.

Holmes, A. 1944. *Principles of Physical Geology.* 2nd revised edition 1965. London: Nelson.

Homère. *L'Iliade.* II, 575 et VIII, 203.

Jayaputra, A. A. 1984. The temple of Borobudur. *Interdisciplinary Sympos. University of Naples.* 9-10 April.

Jelgersma, S. 1979. Sea-level changes in the North-Sea basin. In the Quaternary History of the North-Sea 233-248 *Acta Univ. Ups. Symp.* Uppsala.

Laronde, A. 1981. Variations du niveau de la mer sur les côtes de la Cyrenaïque à l'époque historique. *Histoire et archéologie: Ports et villes englouties.* Ser. 50.

Maheu, R. 1973. Appel pour la sauvegarde de Mohenjo Daro. *Courrier UNESCO,* Dec.

Muir, J. 1967. *Gentle Wilderness: the Sierra Nevada.* New York: David Brower, Ballantine Books.

Muller, L. 1963. Die Standfestigkeit von Felsboschungen als speikisch geomechanische. *Aufgabe Rock Mech. and Eng. Geol.* I, 1; 50-71.

Netherlands Commemorative Volume produced by the Netherlands Member Society (1985) in honour of the *50th ann. of the Int. Soc. Soil Mech. Fdb. Engng.* Rotterdam: de Leeuw.

Paskoff, R., P. Trousset & R. Dalongeville. 1981. Variations relatives au niveau de la mer en Tunisie depuis l'Antiquité. *Histoire et Archéologie: Ports et Villes englouties.* Ser. 50.

Potiche (de) 1891. *La baie du Mont Saint-Michel et ses approches.* Paris.

Raikes, R. L. & R. H. Dyson. 1961. The Prehistoric Climates of Baluchistan and the Indus Valley, *Am. Anthrop.,* 63: 265-281.

Raikes, R. L. 1964. The end of the ancient cities of the Indus, *Am. Anthrop.,* 66: 284-299.

Rau, J. L. & P. Nutalaya. 1983. Geology of the Bangkok Clay. *Geol. Soc. Malaysia* 16: 99-116.

Schmiedt, G. G. 1981. Les viviers romains de la côte tyrrhénienne. *Histoire et Archéologie* 50.

Seed, H. B. 1979. See Chapter 1.

Silverberg, R. *Ciudades perdidas y civilizaciones desaparecidas.* Mexico: Diana. 118-120.

Suarès, A. 1910. *Le voyage d'un condottiere.* Paris: Cornély.

Taine, H. 1860. *Voyage en Italie.* Paris: Hachette.

Wallace, R. E., (1968) Notes on stream channels offset by the San Andreas Fault, Southern Coast Ranges. California. *Proc. of Conf. on Geologic Problems of San Andreas System. Stanford Univ. Publ. XI.*

Wallace, R. E. San Andreas fault. *Dept. of Interior Geographical Survey.*

West, R. G. 1968. *Pleistocene geology and biology.* London: Longmans.

West, R. G. 1972. Relative land sea-level changes in southeastern England during the pleistocene. *Philos. Trans R. Soc. London.* 272 87-98.

Subject and author index